THE 2ND EDITION
MATERIALS
PHYSICS
COMPANION

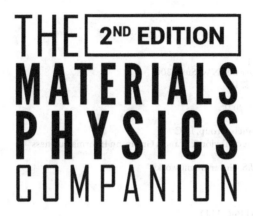

THE 2ND EDITION MATERIALS PHYSICS COMPANION

ANTHONY C. FISCHER-CRIPPS

Fischer-Cripps Laboratories Pty Ltd
Sydney, Australia

CRC Press
Taylor & Francis Group
Boca Raton London New York

CRC Press is an imprint of the
Taylor & Francis Group, an **informa** business

Cover design by Ray Cripps.

CRC Press
Taylor & Francis Group
6000 Broken Sound Parkway NW, Suite 300
Boca Raton, FL 33487-2742

First issued in hardback 2021

© 2015 by Taylor & Francis Group, LLC
CRC Press is an imprint of Taylor & Francis Group, an Informa business

No claim to original U.S. Government works

Version Date: 20140709

ISBN 13: 978-1-138-44146-0 (hbk)
ISBN 13: 978-1-4665-1782-0 (pbk)

Visit the Taylor & Francis Web site at
http://www.taylorandfrancis.com

and the CRC Press Web site at
http://www.crcpress.com

This book is dedicated to the late
Robert Winston Cheary.
Cheary by name, cheery by nature.

Contents

Preface

This book aims to present the minimum of "what you need to know" about materials physics in a semi-introductory manner. The book is a companion to more detailed books in solid state and materials physics. I hope that the book will provide answers to some difficult questions faced by undergraduate students of physics, as well as serve as a handy reminder for professional scientists who need to have just a brief refreshment of a particular subject area studied long ago.

Those readers who have had a science education at the University of Technology, Sydney, will recognise the hands of Geoff Anstis, Bob Cheary, Walter Kalceff, Les Kirkup, John Milledge, Tony Moon, Geoff Smith and Ray Woolcott in this book. To them, my former teachers, I express my gratitude for a very fine education in materials physics, even if at the time I did not appreciate their efforts. As well, I was greatly assisted by the knowledgeable and helpful staff at the CSIRO Division of Industrial Physics, especially Howard Lovatt and Karl-Heinz Muller, who answered my many questions with kindness and authority.

Finally, I thank Tom Spicer for his sponsorship of the first edition at the Institute of Physics Publishing, and John Navas and Francesca McGowan at Taylor & Francis for their continued support for this second edition.

Tony Fischer-Cripps,
Killarney Heights, Australia

Part 1

Introduction to Materials Physics

1.1 Crystallography

Summary

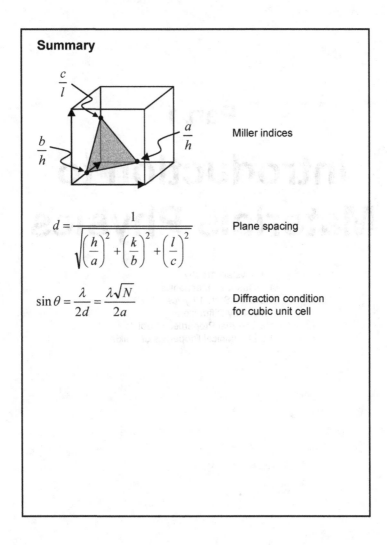

Miller indices

$$d = \cfrac{1}{\sqrt{\left(\dfrac{h}{a}\right)^2 + \left(\dfrac{k}{b}\right)^2 + \left(\dfrac{l}{c}\right)^2}}$$

Plane spacing

$$\sin\theta = \frac{\lambda}{2d} = \frac{\lambda\sqrt{N}}{2a}$$

Diffraction condition
for cubic unit cell

1.1.1 Crystallography

Crystallography is the science of crystals. **Crystals** are solids whose atoms are arranged in a regular repeating pattern that extends through the solid over large distances. Examples are: diamond, sodium chloride, graphite. Since a crystal has to have a regular repeating pattern of atoms within it, it is therefore a term which can only apply to a solid (whose long-range atomic arrangement is static). Crystals may even be made to form in the laboratory from biological structures like DNA molecules. In aome cases, materials for large single crystals (e.g., graphite) while others contain a large number of crystals arranged in random orientation and are called **polycrystalline**.

Solids which are not crystalline are called **Amorphous solids.** They have no long-range regular repeating pattern of atoms or molecules. Examples are glass and most plastics. In these materials, there is an orderly structure in the neighbourhood of any one atom, but these structures themselves are tangled together and there is no long range order that is regularly repeated throughout the material.

The science of crystallography concerns the measurement and the nature of the repeating pattern of atoms. That is, the distance between each atom and the types of atom comprising the crystal. This is usually done by directing a beam of **x-rays** onto the surface of the solid and examining how the beam is scattered. Not only x-rays, but also **neutrons** and **electrons** can be used to determine crystalline properties.

Of course one could in principle establish the nature of a crystal by just looking at it with a microscope, but an ordinary visible light microscope cannot be used because the wavelength of visible light, in the 100's of nm range, is too large to resolve the individual placement of atoms in a crystal (usually of the order of less than 1 nm). This is why shorter wavelength x-rays are usually used. But, x-rays cannot be focussed to form an image like visible light, and so the scattered, or diffracted, beam of x-rays has to be analysed and the position of the atoms inferred from the angle at which constructive and destructive interference of the scattered beam has occurred. This intereference occurs because the atoms of the crystal act like the regular spacing of a diffraction grating.

It has been largely through the measurement of the properties of crystals that our knowledge of the physics of the solid state has arisen.

1.1.2 Lattice

The atoms that comprise a solid are generally arranged in an ordered crystalline state. A crystal is a solid in which the atoms are arranged in such a way as to be periodic. The most basic structure associated with this periodic geometry is a mathematical construction called a **crystal lattice** or a **space lattice**.

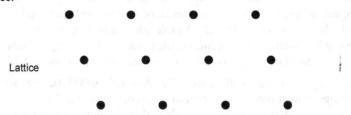

Lattice

A crystal lattice is a set of points in space at which the atomic arrangement of atoms is the same no matter which point is chosen. In the simplest case, the **lattice points** are identical to the atom positions (when all the atoms are of the same type). This type of lattice is called a **Bravais lattice**. However, crystals in solids can consist of a mixture of different types of atoms. The concept of a lattice is still appropriate, although in this case, the lattice points may not correspond to the location of a particular type of atom. Lattice points are points in space at which the atomic arrangement is identical in any one particular direction. Alternatively, we can say that when one translates one's position from one lattice point to another, the arrangement of atoms remains unchanged.

Lattice
+
Basis
=
Crystal structure

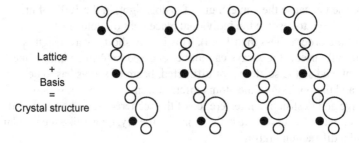

The repeating array of atoms is called the **basis**. The basis, when superimposed upon the crystal lattice, provides a mathematical framework for a description of the **crystal structure** of the solid. A lattice with a basis is a **non-Bravais lattice**, but can be shown to consist of a combination of Bravais lattices for each individual atomic type.

1.1.3 Unit Cell

A lattice may be thought of as consisting of a large number of repeating groups of atoms which is called a unit cell. A unit cell is defined by **lattice vectors a, b** and **c** which begin on lattice points.

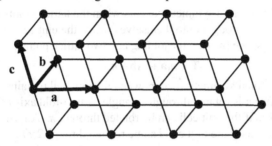

A translation in space of the crystal lattice is written mathematically as:

$$\mathbf{T} = n_1\mathbf{a} + n_2\mathbf{b} + n_3\mathbf{c} \quad \text{where } n_1, n_2 \text{ and } n_3 \text{ are integers.}$$

The volume of a unit cell is calculated from: $V = (\mathbf{a} \times \mathbf{b}) \bullet \mathbf{c}$

The choice of vectors which define a unit cell is not unique. A **primitive unit cell** is one which gives the minimum volume, but is not always the most convenient or illustrative. Consider two representations of a face-centred cubic unit cell:

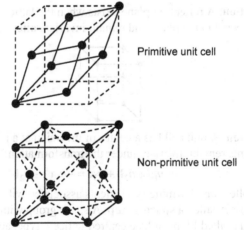

Primitive unit cell

Non-primitive unit cell

In this case, the non-primitive unit cell is more visually appealing and symmetrical and although containing more elements, provides more useful information about the physics of the structure than does the primitive unit cell.

1.1.4 Symmetry

Symmetry operations on a lattice are those which bring the lattice points onto themselves after a translation, rotation, etc. There are several different types of symmetry properties.

Translation: Translation symmetry occurs when the lattice points remain invariant, or are brought onto themselves, when the unit cell is translated in space by a vector joining any two lattice points.

$$\mathbf{T} = n_1\mathbf{a} + n_2\mathbf{b} + n_3\mathbf{c}$$

Rotation: Rotational symmetry occurs when the unit cell remains invariant after it is rotated through an angle. A two-fold axis of rotation exists if the unit cell can be rotated through π. An n-fold axis of rotation exists if the unit cell can be rotated through $2\pi/n$.

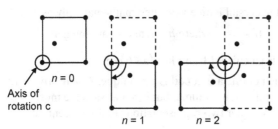

$n = 0$

Axis of rotation c

$n = 1$ $n = 2$

In two fold symmetry, for an atom located at x, y, z there must be an identical atom located at –x, – y, z for a rotation around the c axis.

$$\begin{pmatrix} -1 & 0 & 0 \\ 0 & -1 & 0 \\ 0 & 0 & 1 \end{pmatrix} \begin{pmatrix} x \\ y \\ z \end{pmatrix} = \begin{pmatrix} \overline{x} \\ \overline{y} \\ z \end{pmatrix}$$

Reflection: A reflection plane exists when the lattice points are mirror reflected in the plane and remain invariant after reflection.

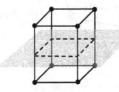

Inversion: A unit cell has a centre of inversion at a point when the lattice points remain invariant under the transformation:

$$\mathbf{R} = n_1\mathbf{a} + n_2\mathbf{b} + n_3\mathbf{c} = -n_1\mathbf{a} + -n_2\mathbf{b} + -n_3\mathbf{c}$$

The collection of symmetry elements associated with a particular unit cell structure is called a **space group**. A space group with a body centred lattice is called I type, a face centred lattice F type, and a primitive lattice P type. Example: An I type lattice with four-fold symmetry around the **c** axis and two fold symmetry around the **a** and **b** axes is written I422 and is the tetragonal lattice. Space groups are used to identify crystal systems from x-ray diffraction data.

1.1.5 Bravais Lattice

For geometrical reasons, there are only 14 types of space lattices that satisfy symmetry operations such as translation, rotation, reflection and inversion. Each of these 14 lattices is called a **Bravais lattice**. There are seven convenient **crystal systems** in the set of Bravais lattices: cubic, tetragonal, orthorhombic, trigonal, monoclinic, hexagonal and triclinic.

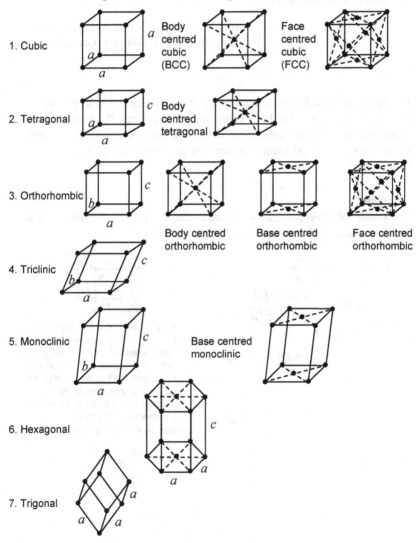

1.1.6 Lattice Parameters

Physical properties of materials are related to information about the crystal structure.

mass per unit cell

$$\text{density } \rho = \frac{m}{V}$$

volume per unit cell

For a cubic structure, $V = a^3$

$$m = \frac{rM}{N_A}$$ where r is the number of molecules per unit cell and N_A is Avogadro's number.

$$\rho = \frac{rM}{N_A a^3}$$

In a face centred cubic structure, the number of molecular units per unit cell is a multiple of 4. For body centred cubic structures, the number of molecular units per unit cell is 2, 4, 6, 8... $2n$. Information about the **lattice parameter** a can most readily be determined by x-ray diffraction experiments.

Example

For a molecular weight of 58.45 g, and a density of 2.15×10^3 kg m^{-3}, and an FCC structure, determine the lattice parameter a.

$$\rho = \frac{rM}{N_A a^3}$$

$$2.15 \times 10^3 = \frac{0.05845(r)}{6.023 \times 10^{23} a^3}$$

$$r = 4$$

$$\therefore a = 5.652 \times 10^{-10} \text{ m}$$

Number of molecules per unit cell: Each of the eight corner atoms contributes 1/8 to the mass of the unit cell because it is shared with eight other unit cells. Each of the six face atoms contributes ½ to the mass of the unit cell because it is shared with one other unit cell, so in this example,

$$r = 8\left(\frac{1}{8}\right) + 6\left(\frac{1}{2}\right)$$

$$= 4$$

1.1.7 Miller Indices

Any three points in a unit cell define a plane in space within the cell. Planes within a unit cell are important for x-ray diffraction. Miller indices are a way of defining a particular plane in a unit cell.

An (*hkl*) plane intersects each of the axes of a unit cell at *a*/*h*, *b*/*k* and *c*/*l*.

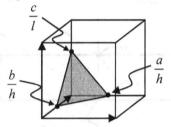

The set of *hkl* numbers when expressed as the smallest possible integers is called the **Miller indices** of the plane. To find the Miller indices, the intercepts of the axes are expressed in terms of the lattice parameters, inverted, and reduced to the lowest possible integers.

Some examples of (*hkl*) planes in a cubic unit cell.

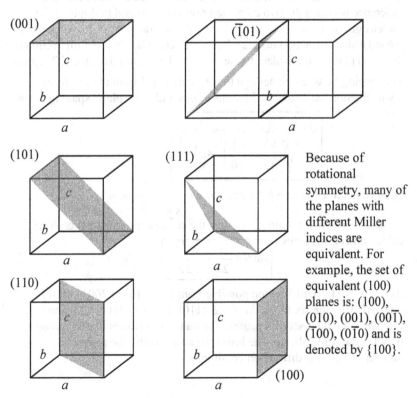

Because of rotational symmetry, many of the planes with different Miller indices are equivalent. For example, the set of equivalent (100) planes is: (100), (010), (001), (00$\bar{1}$), ($\bar{1}$00), (0$\bar{1}$0) and is denoted by {100}.

1.1.8 X-Ray Diffraction

X-ray diffraction is the most important and direct method of determining the properties of crystals. The condition for **constructive interference** for x-rays reflecting from two parallel planes in a crystal is determined by **Bragg's law**:

$$n\lambda = 2d\sin\theta$$

where d is the spacing between the planes.

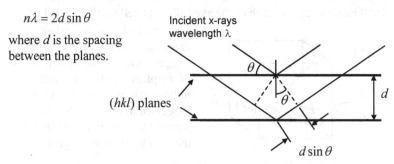

Bragg's law states that constructive interference occurs when the path difference between the two diffracted rays is an integral multiple of the wavelength of the incident x-ray. First order diffraction occurs when $n = 1$, second order diffraction occurs when $n = 2$, etc. The 2nd order diffraction from the (111) plane is equivalent to the 1st order diffraction from the (222) plane.

The spacing between planes can thus be calculated from the results of x-ray diffraction experiments. In terms of h, k and l, the **plane spacing** d can be expressed:

$$d = \frac{1}{\sqrt{\left(\dfrac{h}{a}\right)^2 + \left(\dfrac{k}{b}\right)^2 + \left(\dfrac{l}{c}\right)^2}}$$

For a cubic unit cell, $a = b = c$ and so: $d = \dfrac{1}{\sqrt{h^2 + k^2 + l^2}}$

Since $h^2 + k^2 + l^2$ is an integer N, then the condition for diffraction for a cubic unit cell becomes:

$$\sin\theta = \frac{\lambda}{2d} = \frac{\lambda\sqrt{N}}{2a}$$

At $N = 1$, we have diffraction possible from (100), (010), (001), (00$\bar{1}$), ($\bar{1}$00), (0$\bar{1}$0) – the six faces. At $N = 2$, (110), (011), (101)... 12 planes. For $N = 3$, {111} a set of 8 planes. The number of planes in each group is called the **multiplicity**. The lower the multiplicity, the larger the number of possible diffraction conditions.

1.2 Quantum Mechanics

Summary

$$L = m_e vr = \frac{nh}{2\pi}$$ Angular momentum

$$E_n = -\frac{m_e Z^2 q_e^{4}}{8\varepsilon_o^{2} h^2 n^2}$$ Bohr atom

$$\lambda = \frac{h}{mv}$$ de Broglie matter wave

$$-\frac{\hbar^2}{2m}\frac{\partial^2 \Psi}{\partial x^2} + V(x,t)\Psi = i\hbar\frac{\partial \Psi}{\partial t}$$ Schrödinger equation

$$E = \frac{n^2 \pi^2 \hbar^2}{2L^2 m}$$ Infinite square well

$$E_n = \left(n + \frac{1}{2}\right)hf$$ Harmonic oscillator

$$E = -\frac{Z^2 q_e^{4} m}{(4\pi\varepsilon_o)^2 2\hbar^2 n^2}$$ Coulomb potential

$$k = 2\frac{n\pi}{L}$$ Periodic boundary conditions

$$g(E) = \frac{dN(E)}{dE} = \frac{V}{2\pi^2}\left(\frac{2m}{\hbar^2}\right)^{\frac{3}{2}} E^{\frac{1}{2}}$$ Density of states

$$f(E) = \frac{1}{e^{(E-E_F)/kT} + 1}$$ Fermi–Dirac distribution

1.2.1 Bohr Atom

In 1897, **Thomson** demonstrated that cathode rays (observed to be emitted from the cathodes of vacuum tubes) were in fact charged particles which he called electrons. Thomson proposed that the atom consisted of a positively charged sphere in which were embedded negatively charged electrons.

Rutherford subsequently found in 1911 that the electrons orbited at some distance from a central positively charged **nucleus**. Rutherford proposed that electrostatic attraction between the nucleus and the electron was balanced by the centrifugal force arising from the orbital motion. However, if this were the case, then the electrons (being accelerated inwards towards the centre of rotation) would continuously radiate all their energy as **electromagnetic waves** and very quickly fall into the nucleus. $q_e = -1.6 \times 10^{-19}\,\text{C}$

In 1913, **Bohr** postulated two important additions to Rutherford's theory of atomic structure:

1. Electrons can orbit the nucleus in what are called **stationary states** in which no emission of radiation occurs and in which the **angular momentum** L is constrained to have values:

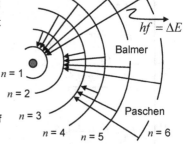

$$L = m_e vr = \frac{nh}{2\pi}$$

The 2π appears because L is expressed in terms of ω rather than f.

2. Electrons can make transitions from one state to another accompanied by the emission or absorption of a single **photon** of energy $E = hf$ thus leading to absorption and emission spectra.

As in the Rutherford atom, the centrifugal force is balanced by Coulomb attraction:

$$\frac{1}{4\pi\varepsilon_0}\frac{q_e^2}{r^2} = \frac{m_e v^2}{r}$$

with the addition that:

$$m_e vr = \frac{nh}{2\pi}$$

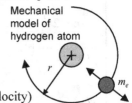

Mechanical model of hydrogen atom

By summing the kinetic energy (from the orbital velocity) and the potential energy from the electrostatic force, the total energy of an electron at a given energy level n is given by:

$$E_n = -\frac{m_e Z^2 q_e^4}{8\varepsilon_0^2 h^2 n^2}$$

Note: $Z = 1$ for the hydrogen atom where the energy of the ground state is –13.6 eV. The energy levels for each state n rises as Z^2. Thus, the energy level of the innermost shell for multi-electron atoms can be several thousand eV.

from which the **Rydberg constant** may be calculated since $\Delta E = hf$

1.2.2 Energy Levels

The stationary states or energy levels allowed by the Bohr model of the atom are observed to consist of sub-levels (evidenced by fine splitting of spectral lines). These groups of sub-levels are conveniently called **electron shells**, and are numbered K, L, M, N, etc., with K being the innermost shell corresponding to $n = 1$. The number n is called the **principal quantum number** and describes how energy is quantised.

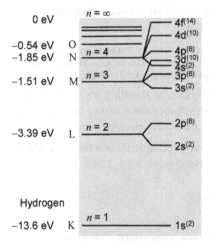

The energy required to move an electron from an electron shell to infinity is called the **ionisation energy**. It is convenient to assign the energy at infinity as being 0 since as an electron moves closer to the nucleus (which is positively charged) its potential to do work is less; thus the energy levels for each shell shown are negative. For hydrogen, the ionisation energy is –13.6 eV. The energies for the higher energy levels are given by:

$$E = -\frac{13.6}{n^2} \quad \text{For hydrogen}$$

The electron-volt is a unit of energy.
1 eV = 1.602×10^{-19} J

At each value of n (i.e., at each energy level) the **angular momentum** can take on several distinct values. The number of values is described by a second quantum number l. The allowed values of l are 0, 1, ... $(n–1)$. Each value of l is indicated by a letter:

A third quantum number m describes the allowable changes in angle of the **angular momentum** vector in the presence of an electric field. It takes the values $–l$ to 0 to $+l$.

A fourth quantum number describes the **spin** of an electron where the spin can be either $–1/2$ or $+1/2$.

$l = 0$	s
$l = 1$	p
$l = 2$	d
$l = 3$	f
$l = 4$	g
$l = 5$	h

According to the **Pauli exclusion principle**, no electron in any one atom can have the same combination of quantum numbers. This provides the basis for the filling of energy levels.

When all the electrons in an atom are in the lowest possible energy levels, the atom is said to be in its **ground state**.

For example, the 3d energy level can hold up to 10 electrons:

$n = 3$

thus: $l = 0, 1, 2 =$ s, p, or d

and: $m = -2, -1, 0, 1, 2$

5 values of m times two for spin thus 10 possible electrons

1.2.3 Matter Waves

The Bohr model of the atom strictly applies only to a single electron orbiting a nucleus and ignores interactions between electrons and other neighbouring atoms. Further, the theory does not offer any explanation as to *why* the angular momentum is to be quantised. Such explanations and treatments can only be explained in terms of **wave mechanics**.

In 1924 **de Broglie** postulated that matter exhibited a dual nature (just as did electromagnetic radiation) and proposed that the wavelength of a particular object of mass m is found from:

$$\lambda = \frac{h}{mv}$$

Because h is a very small number, the wavelength of large objects is very small. For small objects, e.g., electrons, the wavelength is comparable to atomic dimensions.

where mv is the momentum p of the object. The resulting waves are called **matter waves**. In the case of atomic structure, matter waves for electrons are **standing waves** that correspond to particular electron orbits.

For a particular radius r, a standing wave is obtained when the circumference of the path is an integral number of wavelengths: $n\lambda = 2\pi r$.

Thus, from the expression for matter waves, we obtain: $2\pi r = n\left(\dfrac{h}{mv}\right)$

$$mvr = n\frac{h}{2\pi} \longrightarrow \text{Bohr condition for stable state since } L = mvr.$$

The **wave-particle duality** of matter means that, inherently, an electron is neither a wave nor a particle but its motion can be quantified using the mathematical equations appropriate to waves and particles. The wave nature of matter is often interpreted as being one of probabilities. The amplitude of a matter wave represents the probability of finding the associated particle at a particular position.

Since matter is described in terms of a probability, there becomes an inherent limitation in what we can know about the motion and position of a particle such as an electron. The **Heisenberg uncertainty principle** quantifies these uncertainties. For momentum and position, the requirement is:

$$\Delta p \Delta x \geq \frac{h}{2\pi}$$

where Δp and Δx are the uncertainties associated with these quantities. The more we reduce the uncertainty in one, the more the uncertainty in the other increases.

1.2.4 Schrödinger Equation

The total energy of a particle (e.g., an **electron** in an atom, a free electron, an electron in an electric field, a conduction electron in a solid) is the sum of the potential and kinetic energies. Expressed in terms of **momentum** p, and mass m, this is stated:

$$E = \frac{1}{2}mv^2 + V$$

$$= \frac{1}{2}pv + V \quad \text{since} \quad p = mv$$

$$E = \frac{p^2}{2m} + V \quad \text{Energy equation}$$

> let $\omega = 2\pi f$
>
> and $k = \dfrac{2\pi}{\lambda}$
>
> and $\hbar = \dfrac{h}{2\pi}$
>
> thus $E = hf = \hbar\omega$

Considering movement in one dimension only,

let $\hat{p} = -i\hbar \dfrac{\partial}{\partial x}$

$\hat{E} = i\hbar \dfrac{\partial}{\partial t}$

$i^2 = -1$

The $\partial/\partial x$ and $\partial/\partial t$ terms are **differential operators**. For example, when the differential operator $\partial/\partial t$ acts on the displacement variable x, we obtain $\partial x/\partial t$ = velocity. We use the symbol ∂ instead of d here because we will apply these operators to a function which is dependent on both x and t, and so the partial derivatives must be used. The differential operator can also act on a function. For example, if $f(x,t) = 2x + 3t$, then $\partial f(x,t)/\partial t = 3$.

Now, the **potential energy**, V, may depend upon the position and time, and so in general, $V = V(x,t)$.

The quantities $\hat{p}$ and $\hat{E}$ above are **differential operators**. When they operate on a function involving x and t, the result is the momentum p and energy E, respectively. The energy equation becomes a differential operator equation.

$$i\hbar \frac{\partial}{\partial t} = -\frac{\hbar^2}{2m}\frac{\partial^2}{\partial x^2} + V(x,t)$$

Note: $(-i)^2 = -1$

We now let this differential operator equation operate on a function Ψ. Ψ is called a **wave function**, and may itself be a function of x and t.

We can now write the total energy as:

$$-\frac{\hbar^2}{2m}\frac{\partial^2 \Psi}{\partial x^2} + V(x,t)\Psi = i\hbar \frac{\partial \Psi}{\partial t}$$

This is the **Schrödinger equation**. It is a differential equation. The solution to this equation is the function Ψ. Thus, we solve the Schrödinger equation by finding the form of Ψ for various forms of the potential function V.

A simple example is $\Psi(x,t) = A \sin(kx - \omega t)$, the wave function for a sinusoidal travelling wave. When dealing with more complicated functions like matter waves, the wave function is a complex function. For example, one solution of the wave equation (a free electron) has the form: $\Psi(x,t) = (A \cos(kx) + B \sin(kx))e^{-i\omega t}$

1.2.5 Solution to the Schrödinger Equation

The solution to the Schrödinger wave equation is the **wave function** Ψ. For many cases of interest, the potential function V is a function of x only, that is, the potential is static (independent of time). This allows the wave equation to be separated into time-independent and time-dependent equations that can be readily solved independently.

$$-\frac{\hbar^2}{2m}\frac{\partial^2 \Psi}{\partial x^2}+V\Psi = i\hbar\frac{\partial \Psi}{\partial t}$$

let $\Psi = \psi(x)\phi(t)$

Schrödinger
differential wave
equation

$$-\frac{\hbar^2}{2m}\frac{\partial^2 \psi\phi}{\partial x^2}+V\psi\phi = i\hbar\frac{\partial \psi\phi}{\partial t}$$

$$-\frac{\hbar^2}{2m}\frac{\partial^2 \psi\phi}{\partial x^2}\frac{1}{\psi\phi}+V = i\hbar\frac{\partial \psi\phi}{\partial t}\frac{1}{\psi\phi}$$

ϕ is a function of t
ψ is a function of x

$$\frac{1}{\psi}\left[-\frac{\hbar^2}{2m}\frac{\partial^2 \psi}{\partial x^2}+V\psi\right] = i\hbar\frac{\partial \phi}{\partial t}\frac{1}{\phi}$$

$$-\frac{\hbar^2}{2m}\frac{\partial^2 \psi}{\partial x^2}+V\psi = G\psi$$

Time-independent
Schrödinger
equation

$$i\hbar\frac{\partial \phi}{\partial t}\frac{1}{\phi} = G$$

Time-dependent
Schrödinger
equation

G is a constant that just connects the two equations. It is termed the **separation constant** because it allows the variables to be separated into two equations. The physical significance of G will be shown to be the energy E.

The resulting solutions of these equations are functions, one a function of x, the other a function of t. When these two functions are multiplied together, we obtain the **wave function**. The wave function $\Psi(x,t)$ is the solution to the original Schrödinger differential wave equation.

$$\boxed{\Psi(x,t) = \psi(x)\phi(t)}$$

$\psi(x)$ is a solution to the **time-independent equation**. $\phi(t)$ is the solution to the **time-dependent equation**.

In general, there may be many solutions to the time-independent equation, each differing by a multiplicative constant. The collection of functions $\psi(x)$ that are solutions are called **eigenfunctions**, or **characteristic functions**. The **eigenfunctions** for the time-independent equation $\psi(x)$ determine the space dependence of the wave function Ψ. The quantum state associated with an eigenfunction is called an **eigenstate**.

1.2.6 Interpretation of the Wave Function

We might ask, just what is the physical significance of the **wave function** Ψ? In classical physics, a wave function describes the displacement of a particle, or the amplitude of an electric field, or some other phenomenon, at a given coordinate x at some time t. For example, the amplitude of the electric field in an electromagnetic wave at some location x at time t can be expressed:

$$E(x,t) = E_0 \sin(kx - \omega t)$$

The **energy density** is the energy contained within the electric field per unit volume (J m^{-3}). The **intensity** of the field is a measurement of power (i.e., rate of energy transfer) transmitted per unit area by the field. The average (or rms) power of
an electromagnetic wave is: $I_{av} = \varepsilon_o c \left[\frac{1}{2} E_o^{\ 2} \right]$

The important feature here is that the energy carried by a wave is proportional to the *square* of the amplitude of the wave. In the case of electromagnetic waves, what we are really measuring as energy is the density of photons (i.e., number per unit volume) - since each photon carries its own quanta of energy.

By analogy to the case of photons, the wave function for an electron has a connection with the energy carried by it since the **Schrödinger equation** is an energy equation. **Born** postulated that the square of the amplitude of the wave function is a measure of the **probability density** of the electron. Since Ψ is complex, in order to obtain a real physical value for this probability, we use the product:

$$\boxed{P(x,t) = |\Psi|^2 = \Psi^* \Psi}$$

where Ψ^* is the complex conjugate of Ψ. $|\Psi|^2$ is interpreted as a **probability density function**. For example, the probability that an electron is located within a small increment Δx around x at time t is $P(x,t)\Delta x$.

When a small particle, such as an electron (or a proton, or a photon), travels from place to place, it does so using all possible paths that connect the two places. Some paths are more probable than others. The electron is not smeared out into some kind of wave – it retains its particle-like nature. It is the *probabilities* that are wave-like.

For example, say there is a spate of car thefts in the east part of a city. The next week, more than the usual number of thefts occur in the centre of the city. In the next week, it is found that a large number of thefts occur in the west. A probability wave is moving from east to west through the city – whereby the chances of finding an increased number of car thefts depend upon the time (i.e., which week) and the place (east, centre or west).

1.2.7 The Time-Dependent Equation

We shall use the example of an electron in a potential field to illustrate the nature of quantum mechanics, although it should be remembered that the principles also apply to other objects such as protons, neutrons and photons.

The solution to the time-dependent equation involves the use of an **auxiliary equation**. We proceed as follows:

$$i\hbar \frac{\partial \phi}{\partial t} \frac{1}{\phi} = G$$

$$\frac{\partial \phi}{\partial t} = -G\phi \frac{i}{\hbar}$$

Auxiliary equation $m = -G\dfrac{i}{\hbar}$

$$\therefore \phi = e^{-\frac{Gi}{\hbar}t}$$

$$= \cos\frac{Gt}{\hbar} - i\sin\frac{Gt}{\hbar}$$

Euler's formula
$$e^{(a+bi)x} = e^{ax}\left(\cos bx + i\sin bx\right)$$

i.e., $\phi(t)$ has frequency $\omega = \dfrac{G}{\hbar}$

but $\omega = \dfrac{E}{\hbar}$ where E is the total **energy** of the particle.

$$\therefore G = E$$

$$\phi(t) = e^{-i\frac{Et}{\hbar}}$$

Comparing with the general exponential form of the **wave equation**:

$$y(t) = Ae^{i\omega t}$$

We see that the time-dependent part of the wave function represents the **phase** of the probability wave and so the time-independent part represents the **amplitude** of the wave.

$$\Psi(x,t) = \psi(x)\phi(t)$$

That is, although the total wave function Ψ is a function of x and t, the amplitude of the wave function is independent of t. That is, the positional probability density is independent of time. Under these conditions, the electron is said to be in a **stationary state**. In this case, the probability amplitude is given by:

$$P(x) = |\Psi|^2 = |\psi|^2$$

1.2.8 Normalisation and Expectation

In order for the positional probability of an electron to have physical meaning, the electron must be somewhere within the range between $-\infty$ and $+\infty$. That is:

$$\int_{-\infty}^{\infty} \Psi^*\Psi dx = 1$$

The amplitude of the wave function is found from the solution to the time-independent equation. We shall see that the general solution $\psi(x)$ to this equation contains a constant of arbitrary value. For a particular solution, the value of the constant depends upon the boundary conditions of the problem. The most general situation is that the electron must be somewhere. That is, the total probability of finding the electron between $-\infty$ and $+\infty$ is 1. When the value of the constant has been found from the boundary conditions, then the wave function is said to be **normalised**.

What then is the expected location of a particle, such as an electron? Since the electron must be somewhere between $-\infty$ and $+\infty$, the expected value is the weighted sum of the individual probabilities over all values of x.

$$\langle x \rangle = \int_{-\infty}^{\infty} xP(x,t)dx = \int_{-\infty}^{\infty} \Psi^* x\Psi dx$$

$\langle x \rangle$ is the **expectation value** of the electron's position. This is not necessarily the most likely value of x. The most likely value of x for a given measurement is given by the maximum (or maxima) in the probability density function. The expectation value is the average value of x that would be obtained if x were measured many times. When the probability density is symmetric about $x = 0$, the expectation value for $x = 0$.

Expectation values for **energy** and **momentum** may also be calculated from:

$$\langle E \rangle = \int_{-\infty}^{\infty} \Psi^* \hat{E}\Psi dx$$

$$\langle p \rangle = \int_{-\infty}^{\infty} \Psi^* \hat{p}\Psi dx$$

The terms inside the integral sign are customarily written in the order shown here to reflect a style of "**bra-ket**" operator notation introduced by **Dirac**.

1.2.9 The Zero Potential

Note: Here, we are using the technique of separation of variables with complex roots to determine the general solution to the wave equation.

Consider the case $V(x) = 0$

$$-\frac{\hbar^2}{2m}\frac{\partial^2 \psi}{\partial x^2} = E\psi$$

$$\frac{\partial^2 \psi}{\partial x^2} = \frac{2Em}{-\hbar^2}\psi$$

$$\frac{\partial^2 \psi}{\partial x^2} + \frac{2Em}{-\hbar^2}\psi = 0$$

Euler's formula
$$e^{ix} = \cos x + i\sin x$$

Auxiliary equation: $m^2 + \frac{2Em}{\hbar^2} = 0$ and so $m = \pm i\frac{\sqrt{2Em}}{\hbar}$

letting $k = \frac{\sqrt{2Em}}{\hbar}$ Note that E (and k) can take on any value > 0. The boundary conditions do not require any discreteness (n).

we obtain: $\psi(x) = C_1 e^{+ikx} + C_2 e^{-ikx}$

Converting to trigonometric form using Euler's formula:

$$\psi(x) = C_1 \cos kx + C_1 i\sin kx + C_2 \cos(-kx) + C_2 i\sin(-kx)$$
$$= (C_1 + C_2)\cos kx + (C_1 - C_2)i\sin kx$$

The **eigenfunctions** become: $\psi(x) = A\cos kx + Bi\sin kx$ $\begin{cases} A = C_1 + C_2 \\ B = C_1 - C_2 \end{cases}$

The **wave function** is thus:

$$\Psi(x,t) = \psi(x)\phi(t)$$
$$= (A\cos kx + Bi\sin kx)e^{-i\omega t} \quad \text{since } \frac{E}{\hbar} = \omega$$

or $\Psi(x,t) = (C_1 e^{ikx} + C_2 e^{-ikx})e^{-i\omega t}$

$$= C_1 e^{i(kx-\omega t)} + C_2 e^{-i(kx+\omega t)} \quad \text{in exponential form.}$$

wave travelling $+x$ \qquad wave travelling $-x$

This is a general solution that describes the **superposition** of a wave travelling to the right ($+kx$) and one travelling to the left ($-kx$) with amplitudes C_1 and C_2, respectively.

Note, we need to select one of the possible solutions since we can't have a free electron travelling to the left and then to the right. It has to be one or the other. If it changed direction, then some force would act upon it and it would not be "free."

A particular solution for the case of a probability wave travelling in the $+x$ direction can be found by setting $C_2 = 0$ and so $C_1 = A = B$ and hence:

$$\Psi(x,t) = (A\cos kx + Ai\sin kx)e^{-i\omega t}$$
$$= (Ae^{ikx})e^{-i\omega t}$$

$$\boxed{\Psi(x,t) = Ae^{i(kx-\omega t)} = A\cos(kx - \omega t) + Ai\sin(kx - \omega t)}$$

That this represents a **travelling wave** can be seen from the real part of $\Psi(x, t)$.

$$\Psi(x,t)_{re} = A\cos(kx - \omega t)$$

While the amplitude A of the wave might remain constant, its position is dependent on the time t. That is, whenever $kx - \omega t = \pi/2, 3\pi/2$, etc. we have a node (where the function $\Psi = 0$.) If we fix a time t, then these nodes will appear at periodic intervals of x. As t increases, the positions x for the nodes must also increase and so the nodes "travel" along in the x direction – that is, the wave travels (and the electron travels along with it).

Expectation values:

The expected, or average, value of **momentum** is found from:

$$\langle p \rangle = \int_{-\infty}^{\infty} \Psi^* \hat{p} \Psi dx$$

$$= \int_{-\infty}^{\infty} \Psi^* (-i\hbar)\frac{d\Psi}{dx}dx$$

$$= \int_{-\infty}^{\infty} \Psi^* (-i\hbar)(ik)Ae^{i(kx-\omega t)}dx$$

$$= \int_{-\infty}^{\infty} -i\hbar(ik)\Psi^* \Psi dx$$

$$= \hbar k \int_{-\infty}^{\infty} \Psi^* \Psi dx$$

$$= \hbar k$$

$$\boxed{\langle p \rangle = \sqrt{2Em}}$$ **de Broglie relation**

The expected, or average, value of **energy** E for a free electron is found from:

$$\langle E \rangle = \int_{-\infty}^{\infty} \Psi^* \hat{E} \Psi dx$$

$$= \int_{-\infty}^{\infty} \Psi^* i\hbar \frac{d\Psi}{dt}dx$$

$$= \int_{-\infty}^{\infty} Ae^{-i(kx-\omega t)}i\hbar(-i\omega)Ae^{i(kx-\omega t)}dx$$

$$= \int_{-\infty}^{\infty} -i^2\hbar\omega\Psi^* \Psi dx$$

$$= \hbar\omega$$

$$\boxed{\langle E \rangle = hf}$$ **Planck's equation**

Note that E can take on any value; there is no discreteness (n) in the expression for k (and hence E).

$$\hat{p} = -i\hbar\frac{\partial}{\partial x}$$

$$\hat{E} = i\hbar\frac{\partial}{\partial t}$$

$$k = \frac{\sqrt{2Em}}{\hbar}$$

The positive value of momentum here is consistent with the sign convention we adopted for matter waves, that is, $(kx - \omega t)$ for an electron and wave moving in the positive x direction. If the probability wave is moving in the $+x$ direction, then the electron is very likely also moving in the same direction – since the amplitude (squared) of the probability wave determines where the electron is most likely to be, as if it were carried along by the wave.

Normalisation:

$$1 = \int_{-\infty}^{\infty} \Psi^* \Psi dx = \int_{-\infty}^{\infty} A e^{-i(kx-\omega t)} A e^{i(kx-\omega t)} dx = A^2 \int_{-\infty}^{\infty} dx \quad \text{(Divergent integral)}$$

The difficulty with this normalisation is that the limits of integration are far larger than that which would ordinarily apply in a real physical situation. If the electron is bound by a large, but finite boundary, then (as in the case of a square well potential) a non-zero value of A may be calculated while retaining an approximation to the ideal case of infinite range.

Expectation value of x:

We might well ask what the expectation value is for the position x of the electron.

$$|\Psi|^2 = \Psi^* \Psi = A e^{-i(kx-\omega t)} A e^{i(kx-\omega t)} \qquad |\psi|^2 = \psi^* \psi = A e^{-ikx} A e^{ikx}$$
$$= A^2 \qquad\qquad\qquad\qquad\qquad\text{or}\qquad\qquad = A^2$$

That is, the positional probability density is a constant, which means that the electron has an equal probability of being located anywhere along x. This is in accordance with the uncertainty principle since in this case, the momentum (and hence the velocity) can be precisely calculated, but the position is completely undetermined.

Note that for the equation $\psi(x)$ to be a valid solution to the time-independent part of the Schrödinger equation, it does not matter what the value of k (and hence E) is as long as it is a constant in time and independent of x. This means that for a given value of E, the *square* of the amplitude of the resulting wave function is a constant, independent of x (and t). That is, for a free electron, the electron can be equally likely to be anywhere in x with a velocity v (and momentum p). The uncertainty in x is infinite. The value of p is known precisely. The energy E is not quantised for a completely free electron.

1.2.10 Particle in a Box

A particularly important case which can be solved using the **Schrödinger equation** is that of a one-dimensional motion of an electron between two rigid walls separated by a distance L. Such a scenario is called a **particle in a box**. We wish to compute the probability of finding the electron at any particular position between 0 and L according to the principles of quantum mechanics.

The motion of the electron is assumed to consist of completely elastic collisions between the walls. At any position between the walls, it is assumed to have a constant velocity (and hence, momentum) independent of time. That is, the (kinetic) energy is a constant and expressed as:

$$E = \frac{p^2}{2m}$$

Note, this is NOT simple harmonic motion between the walls. It is *constant velocity* between the walls.

Since the momentum of the electron is a constant, from the **de Broglie** relation $\lambda = h/p$ we have a characteristic single wavelength λ. Therefore, we can expect that the solution to the time-independent wave equation will be of the form of a travelling wave: $\psi(x) = A \sin kx$ where $k = 2\pi/\lambda$.

The boundary conditions associated with the walls are satisfied as long as $k = n\pi/L$, or, the allowed wavelengths of the electron are $2L/n$ where $n = 1,2,3....$ These are **standing wave** patterns between the walls of the box.

The energies for each standing wave are found from the allowed values of momentum and the de Broglie relation:

$$E_n = \frac{n^2 h^2}{8mL^2}$$

$n = 1$ is the **zero point energy**.

Unlike the case of a free electron, the presence of the walls imposes a restriction on the allowed values of E which in turn leads to a non-uniform probability of finding the electron at any particular location between them.

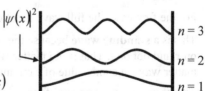

Now,

$$\frac{\partial^2 \psi(x)}{\partial x^2} = -k^2 A \sin kx = -k^2 \psi(x)$$

Thus,

$$\frac{\hbar^2}{2m} \frac{n^2 \pi^2}{L^2} \psi(x) = \frac{n^2 h^2}{8mL^2} \psi(x) = E\psi(x)$$

Probability that the electron will be found within dx of any position x between 0 and L

which is in accordance with the Schrödinger equation:

$$-\frac{\hbar^2}{2m} \frac{\partial^2 \psi}{\partial x^2} = E\psi$$

1.2.11 Infinite Square Well

Consider the case of an **infinite square well potential**:

$$V(x) = \infty \qquad x \le -\frac{a}{2}; x \ge \frac{a}{2}$$

$$= 0 \qquad -\frac{a}{2} < x < \frac{a}{2}$$

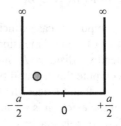

The particle, say an electron, is confined to the region inside the well.

where $V(x) = 0$, we have: $-\dfrac{\hbar^2}{2m}\dfrac{\delta^2\psi}{\delta x^2} = E\psi$ which is the same as the Zero Potential case.

Therefore the **eigenfunctions** are: $\psi(x) = C_1 e^{+ikx} + C_2 e^{-ikx}$.

This is a general solution to the wave equation for the case of $V(x) = 0$, which is the superposition of two travelling waves in opposite directions. In the present case, the electron might be free to travel within the walls of the container, but when it gets to one wall, it bounces back in the other direction. So, here, we do not select one or the other direction as we did for the zero potential case. Instead, both directions must be considered together and further, the travelling waves must have the same amplitude.

In trigonometric form, we obtain (as for the zero potential case):

$$\psi(x) = C_1 \cos kx + C_1 i \sin kx + C_2 \cos(-kx) + C_2 i \sin(-kx)$$
$$= (C_1 + C_2)\cos kx + (C_1 - C_2)i \sin kx$$

However, unlike the case of the zero potential, the travelling waves, in opposite directions, have the same amplitude. Thus, $C_1 = C_2$ and so:

$$\psi(x) = A \cos kx \qquad A = 2C_1$$

Or, alternately, $C_2 = -C_1$ and so:

$$\psi(x) = B i \sin kx \qquad B = 2C_1$$

For the first case, the full wave function is $\Psi(x,t) = (A\cos kx)(e^{-i\omega t})$

This is a **standing wave** because the amplitude term ($A \cos kx$) depends only on x and not on t. That is, no matter what time we look, the amplitude of the matter wave at some value of x remains unchanged.

Therefore, the most general form of solution for the standing wave pattern is the superposition of these two solutions so that:

$$\boxed{\psi(x) = A\cos kx + B i \sin kx} \qquad \begin{cases} A = 2C_1 \\ B = 2C_1 \end{cases}$$

Boundary conditions:

Let $x = \dfrac{a}{2}; \psi(x) = 0$

$$\psi(x) = A\cos\frac{ka}{2} + Bi\sin\frac{ka}{2}$$

$$= 0$$

Let $x = -\dfrac{a}{2}; \psi(x) = 0$

$$\psi(x) = A\cos\frac{-ka}{2} + Bi\sin\frac{-ka}{2}$$

$$= A\cos\frac{ka}{2} - Bi\sin\frac{ka}{2}$$

$$= 0$$

The boundary conditions here are the restriction that the electron has zero probability of being at the walls of the well. This is equivalent to saying that the eigenfunctions reduce to zero at these locations in x. The square of the amplitude of the eigenfunctions is equivalent to the probability density because we are dealing with a **standing wave**, or **stationary state**.

Thus: $A = 0; \; Bi\sin\dfrac{ka}{2} = 0$ or $B = 0; \; A\cos\dfrac{ka}{2} = 0$

Eigenfunctions:

$A = B = 0$ is a trivial solution – that is, the particle is not inside the well. Non-trivial eigenfunctions are found by letting, say, $A = 0$ (or $B = 0$) and letting k take on values such that:

(i) $A = 0; \; Bi\sin\dfrac{ka}{2} = 0$ (ii) $B = 0; \; A\cos\dfrac{ka}{2} = 0$

$$\frac{ka}{2} = n\pi \qquad\qquad\qquad \frac{ka}{2} = n\frac{\pi}{2}$$

$$k = \frac{2n\pi}{a} \quad n = 1,2,3,4... \qquad k = \frac{n\pi}{a} \quad n = 1,3,5,7...$$

$$\text{or}$$

$$= \frac{n\pi}{a} \quad n = 2,4,6...$$

Thus: $\psi_n(x) = A_n \cos k_n x \quad n = 1,3,5,7...$
$$k_n = \frac{n\pi}{a}$$
$\psi_n(x) = B_n i \sin k_n x \quad n = 2,4,6,8...$

Normalisation:

To determine the values of the constants A and B, the eigenfunctions are normalised. For the odd n case:

$$1 = \int_{-a/2}^{a/2} \psi^* \psi \, dx$$

$$= A^2 \int_{-a/2}^{a/2} \cos^2 \frac{n\pi}{a} x \, dx$$

$$= A^2 \int_{-a/2}^{a/2} \frac{1}{2} + \frac{1}{2} \cos \frac{2n\pi}{a} x \, dx$$

$$= A^2 \left[\frac{1}{2} x - \frac{a}{4n\pi} \sin \frac{2n\pi}{a} x \right]_{-a/2}^{a/2}$$

$$= A^2 \left[\frac{a}{2} \right]$$

> Now, $\cos^2 \theta = 1 - \sin^2 \theta$
>
> and $\cos 2\theta = \cos^2 \theta - \sin^2 \theta$
>
> $\sin^2 \theta = \cos^2 \theta - \cos 2\theta$
>
> thus $\cos^2 \theta = 1 - \cos^2 \theta + \cos 2\theta$
>
> $= \dfrac{1 + \cos 2\theta}{2}$

$A^2 = \dfrac{2}{a}$ Similarly for the even case, $B^2 = \dfrac{2}{a}$

The normalised **eigenfunctions** are thus:

$$\psi_n(x) = \sqrt{2/a} \cos k_n x \quad n = 1,3,5,7\ldots$$

$$\psi_n(x) = \sqrt{2/a} \sin k_n x \quad n = 2,4,6,8\ldots$$

while the full **wave function** is written:

$$\Psi_n(x,t) = \left(\sqrt{2/a} \cos k_n x \right)\left(e^{-i\omega t} \right) \quad n = 1,3,5,7\ldots$$

$$\Psi_n(x,t) = \left(\sqrt{2/a} \sin k_n x \right)\left(e^{-i\omega t} \right) \quad n = 2,4,6,8\ldots$$

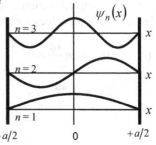

A plot of the probability distributions shows the positional probability of the location of the electron for each of the allowable energy levels. In all cases, the probability is of course zero at the walls where the energy barrier is infinitely high.

Note that because of symmetry around $x = 0$, the **expectation value**, or average value, of x is zero.

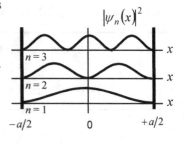

Eigenvalues:

$$k = \frac{n\pi}{a} = \frac{\sqrt{2Em}}{\hbar}$$

$$\frac{n^2\pi^2}{a^2} = \frac{2Em}{\hbar^2}$$

$$\boxed{E = \frac{n^2\pi^2\hbar^2}{2a^2m}} \quad n = 1,2,3,4\ldots$$

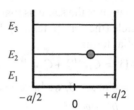

The energy $n = 1$ corresponds to the zero point energy:

$$E_1 = \frac{\pi^2\hbar^2}{2a^2m}$$

Energy is quantised. Each possible value of E is an **eigenvalue**. Unlike the case of a free electron (such as in the zero potential), energy is quantised in this case as a consequence of it being *bound* (between the walls of the well).

When the particle is in the well, the $\Delta x \approx a$. The uncertainty in the momentum is thus:

$$\Delta p = \frac{\hbar}{2\Delta x} = \frac{\hbar}{2a} \quad \text{for } E_1$$

$$p_1 = \sqrt{2mE_1} = \frac{\pi\hbar}{a}$$

$$\Delta p = 2p_1 = \frac{2\pi\hbar}{a}$$

$$\Delta x \Delta p = \frac{2\pi\hbar}{a}a = 2\pi\hbar \quad \textbf{Heisenberg's uncertainty principle}$$

1.2.12 Step Potential

A particularly important potential with practical consequences is the case of a **step potential** V_o. The case of the energy of the electron $E < V_o$ is considered here.

For the case of $x < 0$, the general solution to the wave equation is the same as that developed for a free particle (the zero potential). Expressed in exponential form:

$$\psi(x) = C_1 e^{+ikx} + C_2 e^{-ikx} \quad \text{where } k_1 = \frac{\sqrt{2mE}}{\hbar}$$

$$V(x) = V_o \quad x > 0$$
$$= 0 \quad x < 0$$

For the case of $x > 0$, we have $V(x) = V_o$ and so:

$$-\frac{\hbar^2}{2m}\frac{\partial \psi}{\partial x^2} + V_o \psi = E\psi$$

$$\frac{\partial \psi}{\partial x^2} + \frac{2m}{-\hbar^2}(V_o - E)\psi = 0$$

$$m^2 + \frac{2m}{-\hbar^2}(V_o - E) = 0 \quad \text{auxiliary equation}$$

It is necessary that eigenfunctions ψ must be single-valued, finite and continuous. This enables us to match the two solutions together at the step ($x = 0$).

$$m = \pm \frac{\sqrt{2m(V_o - E)}}{\hbar} \qquad k_2 = \frac{\sqrt{2m(V_o - E)}}{\hbar}$$

$$\psi(x) = C_3 e^{+k_2 x} + C_4 e^{-k_2 x}$$

As x approaches infinity, $\psi(x)$ must be finite and so $C_3 = 0$. At $x = 0$, the value of $\psi(x)$ and $\partial \psi / \partial x$ must match for $x < 0$ and $x > 0$. In evaluating $\psi(x)$ and $\partial \psi / \partial x$ at $x = 0$ we find that:

$$\psi(x) = \frac{C_4}{2}\left(1 + i\frac{k_2}{k_1}\right)e^{ik_1 x} + \frac{C_4}{2}\left(1 - i\frac{k_2}{k_1}\right)e^{-ik_1 x} \quad x < 0$$

$$= C_4 e^{-k_2 x} \quad x > 0$$

That is, unlike the classical Newtonian treatment, quantum mechanics predicts an exponential decrease in the eigenfunction (and also the probability amplitude) on the right hand side of the step. This has important implications for the phenomenon of **tunnelling**.

The first term in the solution for $x < 0$ represents the wave function for the electron approaching the step. The second term represents the electron being reflected from the step. The combined waveforms and associated probability amplitudes are a **standing wave** which represents the probability of the electron being at any point to the left of the step.

1.2.13 Finite Square Well

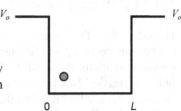

Take, for example, a free electron in a metal. Should an electron near the surface of a metal acquire sufficient kinetic energy to leave the surface, the surface is left with a net positive charge – and so the electron is immediately attracted back towards it. In classical Newtonian mechanics, the electron can only escape the surface completely if it has enough energy to overcome this Coulomb attraction – the **work function** – of the metal. This is an example of a **square well potential**.

The electron is confined to the region inside the well unless it has sufficient energy to overcome the work function V_o.

$$V(x) = V_o \quad x \leq 0; x \geq L$$
$$= 0 \quad 0 < x < L$$

Inside the well, $V(x) = 0$ and so the solution to the zero potential form of the Schrödinger equation can be used:

$$\psi(x) = C_1 e^{ik_1 x} + C_2 e^{-ik_1 x} \text{ where } k_1 = \frac{\sqrt{2mE}}{\hbar}$$

Outside the well, the solution for the step potential can be used:

$$\psi(x) = C_3 e^{k_2 x} + C_4 e^{-k_2 x} \text{ where } k_2 = \frac{\sqrt{2m(V_o - E)}}{\hbar}$$

For a finite solution, $C_3 = 0$ when $x > L$ and $C_4 = 0$ for $x < 0$. The eigenfunctions must also match in slope at the boundary walls of the well (because the **Schrödinger equation** shows that the second derivative of ψ must be finite if $(E - V)$ is finite). The solution to the Schrödinger equation shows that this can only happen at certain values of E.

There are only a finite number of states which can exist where the electron energy is less than V_o. These are called **bound states**. When the electron energy E is greater than this, the electron escapes the bound state of the well and is free (and can have any energy E).

Note that there is a finite probability of the electron being located outside the well even if its energy E is less than V_o.

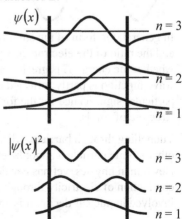

1.2.14 Potential Barrier

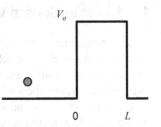

The sides of the finite square well potential V_o can be considered a **potential barrier**. In many physical situations, the width L of the barrier is finite. In classical Newtonian mechanics, an electron can only find itself on the other side of the barrier if it acquires sufficient potential energy to surmount the barrier (i.e., by being given a sufficient amount of an initial kinetic energy, for example, from, say, heating). In quantum mechanics, the solution to the Schrödinger equation for this potential allows for the possibility of the **electron tunnelling** through the barrier and appearing on the other side even when the electron energy is insufficient to surmount the barrier.

On either side of the barrier, the solution to the Schrödinger equation is sinusoidal in accordance with the solution for the zero potential. Within the barrier, (for the case of the electron energy E being less than V_o) the solution is an exponential (as in the step potential). As before, we require the solutions to be continuous and finite for all values of x. In matching the **eigenfunctions** in the three regions $x < 0$, $0 < x < L$ and $x > L$, one possible solution has the form:

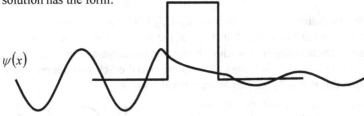

In general, the form of the solution depends upon the width L of the barrier and the ratio of the electron energy to the barrier potential, for example, whether the energy E is greater or less than V_o. For the case of $E > V_o$, the wave function is partly reflected at the barrier, a circumstance which has particular significance for the formation of energy gaps in the band structure of solids.

Tunnelling through barriers, whether we are considering electrons, or any other atomic particle, cannot be explained by conventional classical Newtonian physics, yet has considerable practical importance from the conduction of electricity through contacts and junctions to the processes involved in nuclear decay. It is an everyday occurrence.

1.2.15 Harmonic Oscillator Potential

Consider the case of a **simple harmonic oscillator potential**:

$$V(x) = \frac{1}{2}Cx^2$$

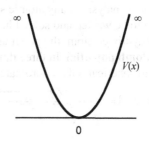

The time-independent equation becomes:

$$-\frac{\hbar^2}{2m}\frac{\partial^2\psi}{\partial x^2} + \frac{Cx^2}{2}\psi = E\psi$$

A **power series** solution (not derived here) yields:

n	Eigenfunctions
0	$\psi_o = A_o e^{-u^2/2}$
1	$\psi_1 = A_1 u e^{-u^2/2}$
2	$\psi_2 = A_2(2u^2 - 1)e^{-u^2/2}$
3	$\psi_3 = A_3(2u^3 - 3u)e^{-u^2/2}$
4	$\psi_4 = A_4(4u^4 - 12u^2 + 3)e^{-u^2/2}$

$$\psi_n(u) = A_n e^{-u^2/2} H_n(u)$$

$\downarrow$

Hermite polynomial

$$\text{where } u = \left(\frac{(Cm)^{1/4}}{\hbar^{1/2}}\right)x$$

The allowed energies, or **eigenvalues**, are:

$$\boxed{E_n = \left(n + \frac{1}{2}\right)hf} \quad n = 0,1,2,3\ldots$$

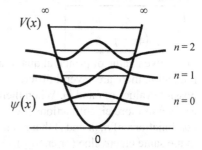

The harmonic oscillator potential has particular importance in describing the state of any system that exhibits small oscillations about a central position (e.g., vibrations of atoms, acoustic and thermal properties of solids, and the response of solids to electromagnetic waves).

Note that compared with the infinite square well potential, the energy levels for the harmonic oscillator are equally spaced hf. Note also the existence of a zero point energy at $n = 0$.

1.2.16 Coulomb Potential – Bohr Atom

A complete analysis of any atomic system will require the Schrödinger equation to be written and solved in three dimensions. For example, for an isolated hydrogen atom, the potential function for the orbiting electron is the **Coulomb potential**. In three dimensions, this is most conveniently written in spherical polar coordinates.

$$V(x,y,z) = \frac{Zq_e}{4\pi\varepsilon_o \sqrt{x^2 + y^2 + z^2}} \longrightarrow \quad q_e = -1.6\times10^{-19}C$$

or $V(r) = -\dfrac{Zq_e^2}{4\pi\varepsilon_o r}$ spherical polar coordinates

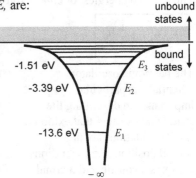

The solution to the time-independent part of the Schrödinger equation has the functional form:

$$\psi(r,\theta,\varphi) = R(r)\Theta(\theta)\Phi(\varphi)$$

The solutions in this case exist only for certain values of the quantum numbers m for $\Phi(\varphi)$, l for $\Theta(\theta)$ and n for $R(r)$ such that:

$$n = 1,2,3...$$

$$l = 0,1,2...n-1$$

$$m = -l,-l+1,...0....+l-1,l$$

The **eigenfunctions** $\psi_{n,l,m}$ provide information about the probability amplitudes, or probability density functions, of the electron for each allowed state.

The **eigenvalues**, or allowed energies E, are:

$$E = -\frac{Z^2 q_e^4 m}{(4\pi\varepsilon_o)^2 2\hbar^2 n^2}$$

For a given value of principal quantum number n, there are several different possible values of l and m. When there are two or more eigenfunctions (i.e., combinations of n, m and l) that result in the same eigenvalue (or energy), these states are said to be **degenerate**.

1.2.17 Superposition

This **superposition** of solutions is a general feature of solutions to linear differential equations. Thus, if the functions $\Psi_n(x, t)$ are solutions to the Schrödinger wave equation, then so is any combination of these.

$$\Psi(x,t) = c_1\Psi_1 + c_2\Psi_2 + ... + c_n\Psi_n$$

The constants c_n allow us to adjust the weighting or proportion of each component wave function if desired.

Linear combinations of eigenfunctions $\psi(x)$ are also solutions to the time-independent equation as long as they correspond to the same value of E.

Consider a *special case* whereby an electron can be in either of one or two separate regions x_1 and x_2. In general, the probability of finding the electron at some position x is given by:

$$P(x,t) = |\Psi(x,t)|^2 = |c_1\Psi_1(x,t) + c_2\Psi_2(x,t)|^2$$

The normalisation condition requires that the total probability, when added over all values of x, is equal to unity so:

$$\int_x P(x,t)dx = \int_x |\Psi|^2 dx = \int_x |c_1\Psi_1 + c_2\Psi_2|^2 dx = \int_x \Psi^*\Psi dx = 1$$

$$= \int_x (c_1\Psi_1{}^* + c_2\Psi_2{}^*)(c_1\Psi_1 + c_2\Psi_2)dx$$

$$1 = c_1{}^2\int_x \Psi_1{}^*\Psi_1 dx + \int_x c_1c_2\Psi_1{}^*\Psi_2 + c_1c_2\Psi_2{}^*\Psi_1 dx + c_2{}^2\int_x \Psi_2{}^*\Psi_2 dx$$

If we take the real part of the wave function, this simplifies to:

$$1 = c_1{}^2 + c_2{}^2 + 2c_1c_2\int_x \Psi_1{}^*\Psi_2 dx$$

The probabilities attached to each wave function individually are:

$$P_1 = c_1{}^2\int_{x_1} \Psi_1{}^*\Psi_1 dx = c_1{}^2 \qquad P_2 = c_2{}^2\int_{x_2} \Psi_2{}^*\Psi_2 dx = c_2{}^2$$

In this special case, each wave function is separately normalised (that is, the electron can only be in either one of two regions x_1 or x_2). The total probability of being in either x_1 or x_2 is thus: $1 = c_1{}^2 + c_2{}^2$.
which means that $2c_1c_2\int_x \Psi_1{}^*\Psi_2 dx = 0$.

This integral, the **overlap integral**, equals zero in this case because we specified that the electron can only be in one region, either x_1 or x_2. This of course is not generally the case. For example, in the double slit experiment there is an appreciable overlap of probabilities from each slit for the electron striking the distant screen midway between the slits.

1.2.18 Transitions

Stationary quantum states occur when the potential V is a function of position only. The complete set of solutions to the Schrödinger equation for a potential $V(x)$ is, by the principle of superposition:

$$\Psi(x,t) = C_1\Psi_1 + C_2\Psi_2 + ... + C_n\Psi_n$$

$$\text{or} \quad \Psi(x,t) = \sum_n C_n(t)\psi_n(x)\phi_n(t) \quad \text{since} \quad \Psi = \psi(x)\phi(t)$$

$$= \sum_n C_n(t)\psi_n(x)e^{-i\frac{E_n}{\hbar}t} \quad \text{and} \quad \phi(t) = e^{-i\frac{Et}{\hbar}}$$

C_n are the weightings for each quantum state n and are expressed as a function of time to take into account the changing probability density of states in a system when transitions occur. Note, the quantum *states* themselves are still stationary, and so we can retain the procedure for the separation of variables, but the probability density of the states may be time dependent. This may happen, for example, during the excitation of an atom where an electron is promoted from a lower energy level, say the ground state, to a higher energy level by the absorption of a photon.

When an electron is in the **ground state**, the probability function is simply:

$$|\Psi|^2 = \Psi^*\Psi = \psi^* e^{+i\frac{E_n}{\hbar}t}\psi e^{-i\frac{E_n}{\hbar}t} = \psi^*\psi$$

That is, $|\Psi|^2$ is independent of time. When an electron makes a transition from state n to m (say the ground state to an excited state), the transition involves a mixed state wave function.

$$\Psi(x,t) = C_n\Psi_n + C_m\Psi_m = C_n(t)\psi_n(x)e^{-i\frac{E_n}{\hbar}t} + C_m(t)\psi_m(x)e^{-i\frac{E_m}{\hbar}t}$$

$$\Psi^*\Psi = \left[C_n\psi_n^* e^{+i\frac{E_n}{\hbar}t} + C_m\psi_m^* e^{+i\frac{E_m}{\hbar}t} \right]\left[C_n\psi_n e^{-i\frac{E_n}{\hbar}t} + C_m\psi_m e^{-i\frac{E_m}{\hbar}t} \right]$$

$$= C_n^2\psi_n^*\psi_n + C_n\psi_n^* C_m\psi_m e^{i\frac{E_n - E_m}{\hbar}t} + C_m\psi_m^* C_n\psi_n e^{-i\frac{E_n - E_m}{\hbar}t} + C_m^2\psi_m^*\psi_m$$

It is sufficient to note from the above that the magnitude of the wave function contains oscillatory terms involving

$$\omega = \frac{E_n - E_m}{\hbar}$$

1.2.19 Atomic and Molecular Potentials

As can be imagined, solving the Schrödinger equation for anything but the very simplest of atoms is not an easy task. The potential function arising from the superposition of many present nuclei and electrons makes even the calculation for two hydrogen atoms interacting in a vacuum a major undertaking.

In physical chemistry, it has been found useful to classify the types of forces between molecules into long and short range forces (using such terms as **Van der Waals, London, dispersion, solvation** forces, etc.). These forces are significant in atoms and molecules which do not normally form ionic, covalent or metallic bonds The forces arise from an instantaneous attraction between electric dipoles between neighbouring atoms/molecules and, in principle, can be described using the Schrödinger equation. In practice, semi-empirical potentials are used to simplify the situation while allowing physical phenomena to be studied in detail.

An example of a widely used pair-potential which describes the interaction between two atoms or molecules is the **Lennard–Jones potential**. This is formed by adding together a long range attractive potential with a short range repulsive potential.

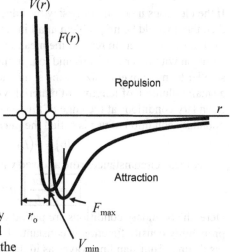

Mathematically, the Lennard–Jones potential is expressed:

$$V(r) = -\frac{A}{r^6} + \frac{B}{r^{12}}$$

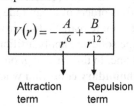

Attraction   Repulsion
term   term

Note that in these potentials, it is customary to assign a potential energy of zero to widely spaced electrons, atoms, ions and molecules and as they approach, the potential becomes more negative. Since we generally assign a positive number to work done *on* a system (energy entering the system) and a negative number to work done *by* a system (energy leaving the system) work has to be done *on* the atoms or molecules to separate them.

1.2.20 Boundary Conditions

There are several interesting consequences of the boundary conditions for the quantum mechanics of an electron in a solid. The basic solution is one of a travelling wave: $\psi(x) = Ae^{ikx} + Be^{-ikx} = A\cos kx + B\sin kx$

In the case of a completely free electron (zero potential), k can have any value and energy is not quantised. When there is a free electron in a potential well, the presence of the walls of the well causes k to only take on certain values according to:

$$k = \frac{n\pi}{L} \quad n = 1,2,3,4\ldots$$

The **fixed boundary conditions** $\psi(x) = 0$ and $\psi(L) = 0$ at walls of the well impose restrictions on the allowable values of k. These boundary conditions result in **standing waves** between the walls. In this situation, the boundary conditions are fixed and the probability density function is periodic in x.

If the electrons in a solid only saw the edges of the solid, and nothing else, then there would be no problem at all in setting $\psi(x) = 0$ and $\psi(L) = 0$ at these locations. However, in reality, the mean free path of electrons (before collision with other electrons and ions in the solid) is orders of magnitude smaller than the overall size of the solid and so it is more realistic to consider a small volume V of length L of material within the solid and apply artificial boundary conditions at the edge of this small volume. In this case, a good boundary condition to use is called the **periodic boundary condition** where:

$$\psi(0) = \psi(L)$$

Under these circumstances, the allowed values of k are:

$$\boxed{k = 2\frac{n\pi}{L}} \quad n = 0, \pm1, \pm2, \pm3, \pm4\ldots$$

Here, the boundary conditions are periodic in x with a period L but the probability density function is constant. The solutions are **travelling waves** (as distinct from standing waves as in the case of the fixed boundary conditions). If we were to take a snapshot of a wave at time t, the nodes of a standing wave stay in the same position no matter what the value of t. The nodes of a travelling wave travel along in the direction of wave motion and so their location depends on the time t.

1.2.21 k Space

We have seen that the solution to the time-independent part of the Schrödinger equation for simple potentials is of the form:

$$\psi(x) = Ae^{ikx} + Be^{-ikx} \quad \text{where} \quad k = \frac{\sqrt{2Em}}{\hbar} = \frac{2\pi}{\lambda}$$

It is convenient to express the energies in terms of k (a **dispersion relation**):

$\lambda = 2\pi\hbar/p$ **de Broglie** wave relationship

$$p^2 = \frac{4\pi^2\hbar^2}{\lambda^2}$$

$$E = \frac{p^2}{2m} = \frac{\hbar^2 4\pi^2}{2m\lambda^2}$$

$$= \frac{\hbar^2}{2m}k^2 \quad \text{where} \quad k = \frac{2\pi}{\lambda} = \frac{p}{\hbar}$$

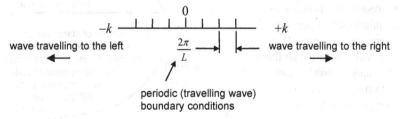

whereupon it can be seen that the dispersion of E is such that $E \propto k^2$. .

The **wave number** k is a vector quantity which characterises the spatial characteristics of the wave (much like the frequency ω characterises the time characteristics of a wave). For the **zero potential**, solutions to the Schrödinger equation can occur at any value of k. However, boundary conditions imposed by the walls of a potential well mean that solutions to the Schrödinger equation occur at certain fixed values of k.

$$k = \frac{2\pi}{\lambda} = \frac{p}{\hbar} = n\frac{\pi}{L} = n\frac{2\pi}{L}$$

standing wave
boundary conditions
$n = 1,2,3,4\ldots$

periodic (travelling wave)
boundary conditions
$n = 0, \pm1, \pm2, \pm3, \pm4\ldots$

Allowed quantum states are uniformly distributed with respect to the value of k. If we were to represent these allowed states graphically, we could write:

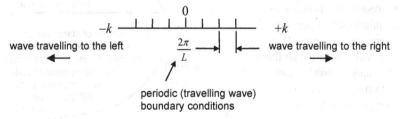

wave travelling to the left

$\dfrac{2\pi}{L}$ wave travelling to the right

periodic (travelling wave)
boundary conditions

1.2.22 Density of States – 1D

Consider a one dimensional, free electron
(**zero potential**) solid of length L.

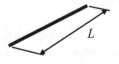

Now, the energy of any particular quantum state is given by $E = \dfrac{\hbar^2}{2m}k^2$.

We wish to determine $N(E)$: how many states exist with
an energy less than a particular value E?

The spacing between each
state in k space is: $\dfrac{2\pi}{L}$

Periodic (travelling
wave) boundary
conditions $\psi(0) = \psi(L)$

$$-k \;\;\underbrace{\!|\,|\,|\,|\,|\,|\,|\,|\,|\,|\,|\,}_{\displaystyle \frac{2\pi}{L}}\;\; +k$$

But, L is the length of the solid in real
space, and so the number of states per unit
length in k space is thus: $\dfrac{1}{2\pi/L} = \dfrac{L}{2\pi}$

We need a factor of two here to
account for states with values of
k on either side of zero.

The number of states less than a particular value of k is thus $2k\left(\dfrac{L}{2\pi}\right)$.

But, since $E = \dfrac{\hbar^2}{2m}k^2$ then $k = \left(\dfrac{2mE}{\hbar^2}\right)^{1/2}$.

The total number of states N less than this value of k is thus:

$$N(E) = \frac{2L}{2\pi}\left(\frac{2mE}{\hbar^2}\right)^{1/2}$$

This is also the number of states with an
energy less than our selected value of E
and is written $N(E)$.

The rate of change of $N(E)$, as E is varied, is the number of states per unit
of energy and is called the **density of states** $g(E)$. It is found from:

$$g(E) = \frac{dN(E)}{dE} = \frac{1}{2}\frac{L}{\pi}\left(\frac{2m}{\hbar^2}\right)^{1/2}E^{-1/2}$$

It is customary to include ×2 for
spin when calculating the
density of states.

$$\boxed{g(E) = \frac{L}{\pi}\left(\frac{2m}{\hbar^2}\right)^{1/2}E^{-1/2}}$$

$N(E)$

The number
of states with energy $<E$
increases as the energy
becomes larger.

E

The density of states $g(E)$
represents the number of
available states per unit of
energy at an energy E. A
high value means that there
are many possible states
within an infinitesimal
energy range $E + dE$.

$g(E)$

The number of states per unit of
energy reduces as the energy
becomes larger. As E becomes
large, the number of states
with energy less than
E still increases but at
a lower rate.

E

1.2.23 Density of States – 2D

Consider a two dimensional, free electron (**zero potential**) solid of length L and area A.

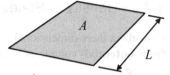

The spacing between each state in k space is: $\dfrac{2\pi}{L}$

Periodic (travelling wave)
boundary conditions $\psi(0) = \psi(L)$

The area per point (i.e., per state) is thus: $\left(\dfrac{2\pi}{L}\right)^2$

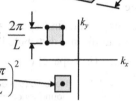

But, L^2 is the area A in real space, and so the number of states per unit area in k space is thus:

$$\frac{1}{(2\pi/L)^2} = \left(\frac{L}{2\pi}\right)^2 = \frac{A}{4\pi^2}$$

Now, each value of k represents a particular energy, so a circle of radius k in k space draws out a contour of a constant value of E. The area of the circle is πk^2.

This is the area in k space, and is not A, the area of the solid in real space.

Thus, the number of states N within the circle is the number of states with an energy less than our selected value of E, that is, $N(E)$.

$$N(E) = \pi k^2 \left(\frac{A}{4\pi^2}\right) = \frac{k^2 A}{4\pi}$$

$$E = \frac{\hbar^2 k^2}{2m}$$

$$k^2 = \frac{2mE}{\hbar^2}$$

→ A factor of ×2 has been applied here to account for spin.

$$N(E) = \frac{2mE}{\hbar^2} \frac{A}{2\pi}$$

If we double E, then there are $N(E)$ twice as many states with energy less than E.

The **density of states** is found from:

$$\boxed{g(E) = \frac{dN(E)}{dE} = \frac{Am}{\pi\hbar^2}}$$

The essential feature in this case is that the density of states for a 2D solid is a constant.

The number of states per unit of energy remains the same no matter which energy we pick.

1.2.24 Density of States – 3D

Consider a three dimensional, free electron (**zero potential**) volume of length L and volume V.

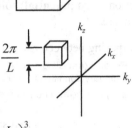

Inside this volume, in k space, allowed states are uniformly distributed with a spacing: $\dfrac{2\pi}{L}$

The volume (in k space) per state is: $\left(\dfrac{2\pi}{L}\right)^3$ — Periodic (travelling wave) boundary conditions $\psi(0)=\psi(L)$

The number of states per unit volume is thus $\left(\dfrac{L}{2\pi}\right)^3$.

But, $L^3 = V$ in real space, and so the states per unit volume in k space can be expressed: $\dfrac{V}{(2\pi)^3}$ where V is the volume in real space.

The surface of a sphere drawn out by a particular value of k represents all the states that have the same energy. Thus, the number of states N within the sphere is the number of states with an energy less than our selected value of E, that is, $N(E)$.

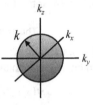

The number of states within the volume is:

$$N(E) = \frac{4}{3}\pi k^3 \frac{V}{(2\pi)^3}$$

but $k^2 = \dfrac{2mE}{\hbar^2}$

thus $k^3 = \left(\dfrac{2mE}{\hbar^2}\right)^{\frac{3}{2}}$

and so $N(E) = \dfrac{4}{3}\pi \dfrac{V}{(2\pi)^3}\left(\dfrac{2mE}{\hbar^2}\right)^{\frac{3}{2}}$

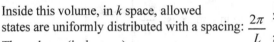

$N(E)$ — The number of states with energy $<E$ increases dramatically as the energy becomes larger.

$g(E)$ — The rate of change of $N(E)$ with E becomes less as the energy becomes larger.

$$g(E) = \frac{dN(E)}{dE} = \frac{4}{3}\pi \frac{V}{(2\pi)^3}\frac{3}{2}\left(\frac{2mE}{\hbar^2}\right)^{\frac{1}{2}}\left(\frac{2m}{\hbar^2}\right)$$

$$\boxed{g(E) = \frac{V}{2\pi^2}\left(\frac{2m}{\hbar^2}\right)^{\frac{3}{2}} E^{\frac{1}{2}}}$$ **density of states**

A factor of ×2 has been applied here to account for spin.

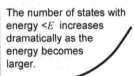

1.2.25 Fermi–Dirac Distribution

The filling of available energy levels in an atom by electrons is governed by the **Pauli exclusion principle** in which it is found that a particular energy level can only accommodate two electrons, one with spin up and the other with spin down.

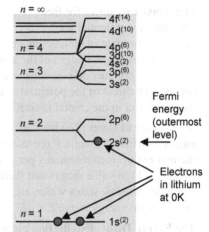

The **Fermi energy** E_f is the energy of the outermost occupied energy level at 0K and is usually of the order of 5 eV for metals. The probability that a particular energy level is full is described by a **probability density function** $f(E)$. At 0K, the probability of finding an electron in an energy level greater than the Fermi energy is zero. The probability of finding an electron in an energy level $< E_f$ is 1.

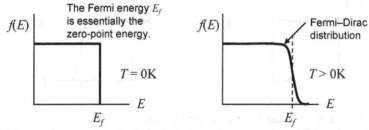

If the temperature is raised, then electrons may acquire additional energy. The thermal energy of electrons at room temperature is ≈ 0.025 eV. Thus, at ordinary temperatures, electrons well below E_f cannot move to a higher energy level because these are occupied already with other electrons. Only electrons with energies close to E_f can increase their energy and jump to unoccupied levels $> E_f$. Thus, some energy levels $< E_f$ become unoccupied and others above E_f become occupied. The distribution of energies for $T > $ 0K is described by the **Fermi–Dirac distribution**:

$$f(E) = \frac{1}{e^{(E-E_F)/kT} + 1}$$

The Fermi–Dirac distribution takes into account the quantised nature of energies of electrons. It approaches the classical Maxwell–Boltzmann distribution (which assumes a continuous distribution of energies) at very high energies.

1.2.26 Electron Energies

Note: For the case of zero potential, it is the presence of the edges that gives rise to allowed values of k and hence quantised energy states.

The **density of states** for free electrons in a potential well describes the state of conduction electrons in a metal bound by the edges of the solid. However, it should be said that treating the conduction electrons as free electrons does not take into account the effect of the potential of the positive ions associated with the atomic nuclei in the crystal lattice.

The density of states $g(E)$ for free electrons within the solid V represents the number of *available states* per unit of energy. A high value means that there are many possible states within an infinitesimal energy range $E + dE$.

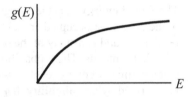

The **Fermi–Dirac distribution** gives the probability of actually finding an electron in an available state as a function of temperature.

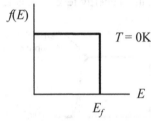

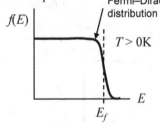

The product of $g(E)$ and $f(E)$ thus gives the energy distribution of the actual electrons present as a function of temperature.

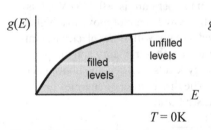

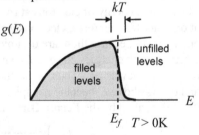

Note that at 0K, the free electrons have a range of energies up to E_f (approx. 5 eV). At temperatures above 0K, only those electrons with energies near to E_f acquire kT of energy. This means that while all free electrons contribute to electrical and thermal conductivity, it is only those near E_f that contribute to the **specific heat**.

1.2.27 Conduction

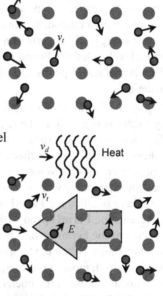

In a conductor, outer shell or **valence electrons** are free to move throughout the crystal lattice; they are not attached to any one particular atom whose nucleus is fixed into the crystal lattice structure. These electrons have a random velocity due to thermal motion and are considered free (although, of course, they still occupy only allowed available energy levels as they travel from atom to atom). Typical velocities of **free electrons** are in the order of $\approx 10^6$ m/s.

When an electric field is applied, the valence, or conduction electrons experience a force and acquire a net velocity over and above their random thermal motion. As they move through the conductor, they suffer collisions with other electrons and fixed atoms and lose velocity and thus some of their kinetic energy. Some of the fixed atoms correspondingly acquire additional **internal energy** (vibrational motion) and the **temperature** of the conductor rises. After collision, electrons are accelerated once more and again by the applied field and suffer more collisions and decelerations and so on. The net **drift velocity** ($\approx 10^{-4}$ m/s) of the electrons constitutes an **electric current**. During collisions, **electrical potential energy** from the voltage source is essentially converted into **heat** within the **conductor**. This is called electrical **resistance**. Increasing the **temperature** increases the random thermal motion of the nuclei of atoms in the crystal structure, thus increasing the chance of collision with a conduction electron and therefore reducing the average drift velocity and increasing the resistance.

The **relaxation time** ($\approx 10^{-14}$ s) is the mean free time between collisions and is independent of the applied field (because the thermal motions are so large in comparison to the drift velocity). The relaxation time also describes how quickly the drift velocity of the electrons ceases upon removal of the field. The **mean free path** of the conduction electrons in a good conductor is ≈ 10 nm. This relatively large value of mean free path compared to the size of atoms is due to the de Broglie wave nature of electrons as they pass through the regular array of atoms in the crystalline lattice.

1.2.28 Fermi Surface for Free Electrons

The outer valence, or conduction electrons in a solid are essentially free particles with a random distribution of velocity. The energy of electrons is kinetic energy. The kinetic energy of the electrons can range from 0 to an upper level called the **Fermi energy**. The Fermi energy is a characteristic of the material and depends upon the concentration of free electrons in the material.

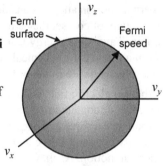

A graph of the range of velocities possessed by electrons in a solid in the x, y and z directions would be a sphere called the **Fermi sphere**. Any point inside this sphere represents a possible velocity (v_x, v_y and v_z) that any electron may have. The outer edge of the sphere is called the **Fermi surface** and represents the **Fermi speed** $\approx 10^6$ m/s. The position of the Fermi surface is independent of temperature (when the temperature of the solid is raised, most of the energy goes into vibrational motion of the nuclei and not that of the valence or conduction electrons).

Fermi sphere

When a field is applied to a conductor, the position of the Fermi surface is no longer centered at $v_x = v_y = v_z = 0$ but is offset by the drift velocity v_d. The offset is extremely small, but it is this offset that is responsible for the macroscopic notion of **electric current**.

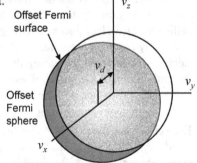

The important issue that should be appreciated here is that it is only those electrons close to the Fermi surface (i.e., those with the greatest velocity and kinetic energy) that are responsible for conduction of electric current. That is, electric current is not due to the **drift velocity** attributed to *all* the free electrons but instead is a result of the net current carried by a relatively small number of electrons whose instantaneous (or thermal) velocity is very close to the edge of the Fermi surface. It can be appreciated, therefore, that any modification to the shape of this surface (such as through the influence of nearby surfaces and the potential of the underlying crystal lattice) will influence the conductivity of the material.

1.3 Solid State Physics

Summary

$$k = \pm\frac{\pi}{a}$$ Brillouin zone

$$E(k) = E_g + \frac{\hbar^2 k^2}{2m_e}$$ Semiconductor: Conduction band

$$E(k) = -\frac{\hbar^2 k^2}{2m_h}$$ Semiconductor: Valence band

Charge carrier concentration

$$n = 2\left(\frac{kT}{2\pi\hbar^2}\right)^{\frac{3}{2}}\left(\frac{m_e}{m_h}\right)^{\frac{3}{4}} e^{-E_g/2kT}$$

1.3.1 Atomic Potentials

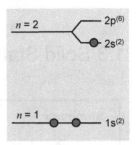

Consider a lithium atom with 3 electrons, 2 of which are in the 1s energy level and the outer valence electron in the 2s energy level. These electrons move in a **potential well**. An example of a potential well is the **infinite square well potential**:

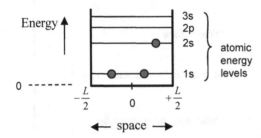

Keep in mind that electrons are not particles in orbit around a nucleus. The energy levels are those associated with the standing wave patterns of probability matter waves.

In the infinite square well potential, the electrons cannot move more than a distance $L/2$ from the centre position. This is a simple potential which can be easily described using the **Schrödinger equation**. A more realistic potential is the **Coulomb potential** and can be represented:

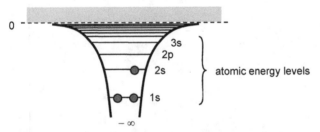

This potential reflects the fact that electrons at higher energy levels are able to occupy more space than those at lower levels. When removed completely from an atom, the electron can occupy any position (i.e., in the diagram above, the zero datum of energy is infinite in the − and + directions) and this is represented by a continuum of states or levels.

1.3.2 Molecules

Staying with a Li atom, let's now consider what happens when we have two atoms close together. We might be first tempted to draw the potentials as:

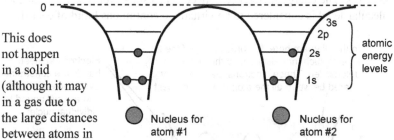

This does
not happen
in a solid
(although it may
in a gas due to
the large distances
between atoms in
gases). In a solid, the electrons in atom #1 are attracted to both its own nucleus and also to some extent by the nucleus of atom #2. For example, the 1s electron orbiting the nucleus of atom #1 is also "owned" to some extent by the nucleus of atom #2. But, due to the **Pauli exclusion principle**, for each atom, we cannot have more than two electrons in the same energy level (i.e., the two 1s electrons for atom #1 as seen by atom #2 are no longer permitted to have energies at the 1s level because there are already two electrons from its own atom at that level).

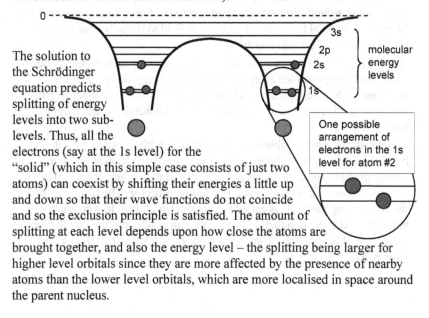

The solution to
the Schrödinger
equation predicts
splitting of energy
levels into two sub-
levels. Thus, all the
electrons (say at the 1s level) for the
"solid" (which in this simple case consists of just two
atoms) can coexist by shifting their energies a little up
and down so that their wave functions do not coincide
and so the exclusion principle is satisfied. The amount of
splitting at each level depends upon how close the atoms are
brought together, and also the energy level – the splitting being larger for higher level orbitals since they are more affected by the presence of nearby atoms than the lower level orbitals, which are more localised in space around the parent nucleus.

1.3.3 Solids

In a real solid, the interacting potentials of many millions of relatively closely spaced atoms cause atomic energy levels to split into a very large number of sub-levels. The energy difference between each sub-level is so fine that each molecular level is considered to be virtually a continuous **band** of energies.

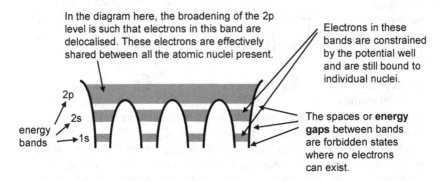

In the diagram here, the broadening of the 2p level is such that electrons in this band are delocalised. These electrons are effectively shared between all the atomic nuclei present.

Electrons in these bands are constrained by the potential well and are still bound to individual nuclei.

2p

2s

energy bands

1s

The spaces or **energy gaps** between bands are forbidden states where no electrons can exist.

If the highest energy band that contains electrons in the ground state (the **valence band**) in a solid is not completely full, *and* the electrons are delocalised, then there are opportunities for electrons within that band to easily move around from state to state *within the band*. Such movement can be readily obtained by applying an electric field to the solid. Such solids are thermal and electrical **conductors**.

If the valence band in a solid is full, and the next highest available band is positioned some distance away in terms of its energy levels, then the electrons within the topmost band cannot easily move from place to place or to the next highest band. Such materials are thermal and electrical **insulators**.

If the next highest available band is positioned fairly close to the valence band, then even at room temperature, there may be sufficient thermal energy given to some electrons to be promoted to this higher level. The material becomes conducting and is a **semiconductor**. The band containing the conducting electrons is called the **conduction band**. In a conductor, the valence band is the conduction band. In a semiconductor, the conduction band (at 0K) is separated from the valence band (defined at 0K) by an **energy gap**.

1.3.4 Energy Bands

The splitting of energy levels into bands is the central feature of the solid state and gives rise to the many varied properties of solids as compared to gases and liquids, whereby the atoms and molecules are widely spaced.

Band width is determined by the separation distance between atoms/molecules. Band also becomes wider at higher energies due to the larger range of movement of electrons in higher energy states.

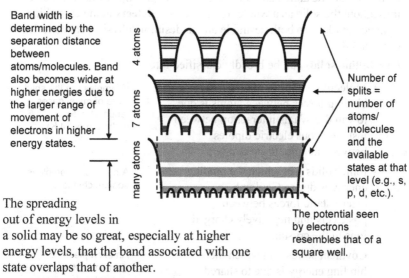

Number of splits = number of atoms/ molecules and the available states at that level (e.g., s, p, d, etc.).

The potential seen by electrons resembles that of a square well.

The spreading out of energy levels in a solid may be so great, especially at higher energy levels, that the band associated with one state overlaps that of another.

Consider the outermost or valence energy level in an atom in a solid in the ground (unexcited) state. In this band, there are N possible sub-levels, each arising from each of the N atoms in the solid. In each of these N sub-levels, there is the possibility of there being two electrons present, each with opposite spin. If a particular atom happens to have two valence electrons, then the valence band for that solid will be completely full – each sub-level being occupied by two electrons. However, if the valence band happens to overlap another empty band, or is very close to an available higher level band, then some electrons can easily move over into that other band, and so what might have been an **insulator** (when considering isolated atoms) becomes a **conductor** (when the atoms are brought together as a solid).

Each individual energy level inside a band represents a particular quantum state. The energy sub-levels inside a band are not equally spaced in terms of energy. That is, the **density of states** is not a constant within an energy band. Further, the distribution of electrons within all the available states varies (e.g., with temperature).

1.3.5 Types of Solids

Crystalline solids are different from **amorphous solids** because their atoms are arranged in a regular, repeating pattern which is called the **crystal lattice**. The repeating unit in the crystal lattice is called a **unit cell** and comprises all the atoms and their relative arrangement, which is repeated throughout the solid as a whole. In most cases, **defects** in the crystalline arrangement lead to both desirable and undesirable physical characteristics of the solid.

Crystalline solids can be broadly classified into:

- Molecular solids – in which the binding forces between atoms is due to van der Waals interaction between instantaneous electric dipoles between neighbouring atoms.

  Examples are the solid state of many gases (such as solid hydrogen at 14 K)

- Ionic solids – in which the binding energy is due to Coulomb electrostatic forces between positively and negatively charged ions in the crystal.

  A common example is sodium chloride.

- Covalent solids – in which the binding energy is due to shared valence electrons between atoms in the solid.

  An example is diamond. Other examples are silicon and germanium.

- Metallic solids – in which valence electrons are effectively shared amongst all the atoms in the solid.

  Metals: good conductors of electricity and heat due to mobility of valence electrons.

Amorphous solids have no long-range regular repeating pattern of atoms or molecules. Examples are glass and most plastics. In these materials, there is an orderly structure in the neighbourhood of any one atom, but this is not regularly repeated throughout the material.

The structure of crystalline solids (such as the spacing between atoms in the unit cell) can be readily studied by the use of **x-ray diffraction** – the orderly arrangement of atoms act as a diffraction grating.

1.3.6 Band Density of States

In a metal, the electrons within the conduction band can (to a first approximation) be considered free within that band. The band itself consists of N closely-spaced energy levels where N is the number of atoms present. The energy of these electrons is proportional to k^2:

upper limit to n

$$E = \frac{\hbar^2}{2m}k^2 \qquad k = 2\frac{n\pi}{L} \quad n = 0, \pm1, \pm2, \pm3, \pm4 \ldots \pm N/2$$

If a is the spacing between the fixed nuclei, and L is the characteristic length of our potential well, then $N = L/a$. Thus, the limits on k for a band containing free electrons is given by:

$$k = 2\frac{n\pi}{L}$$
$$= \pm2\frac{N}{2}\frac{\pi}{L}$$
$$= \pm2\frac{L}{2a}\frac{\pi}{L}$$
$$= \pm\frac{\pi}{a}$$

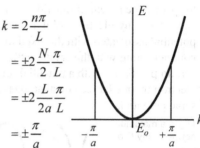

States in k space are evenly spaced by $2\pi/L$. The limits on k are $\pm\pi/a$. A sphere of radius k traces out states with a certain value of E. The number of states dN enclosed within the spheres with radii k and $k+dk$ is proportional to the area dV. The density of states is the number of states per unit energy per unit volume:

$$g(E) = \frac{1}{2\pi^2}\left(\frac{2m}{\hbar^2}\right)^{\frac{3}{2}}E^{\frac{1}{2}}$$

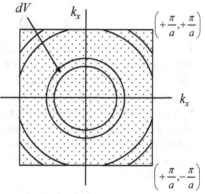

This is appropriate when $k < \pi/a$, but when $k > \pi/a$, the sphere begins to intersect the outer boundary in k space and the number of states per unit energy begins to decrease until the area dV.

There is then an energy gap, followed by the next energy band with its own density of states.

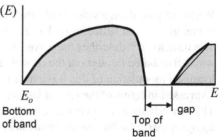

Bottom of band

Top of band

gap

1.3.7 Bloch Function

Previously, we have treated outer level valence and conduction electrons as free, but bounded by a potential well of characteristic size L; which is identified by the size of the unit cell. Now we are in a position to see what effect the presence of the **crystal potential** associated with the fixed nuclei at lattice positions has on these electrons.

The solution of the Schrödinger equation for a particular system requires us to specify the potential function $V(r)$ and this can be quite difficult in the case of a many-electron solid. However, it can be readily accepted that the potential associated with the fixed ions is periodic, with a spatial period commensurate with the crystal lattice. That is, if we have an electron at a certain position r, with a potential function $V(r)$, and then translate the electron by one lattice constant, then the electron will experience the exact same potential.

$V(r)$ is a maximum between any two nuclei.

$V(r)$

$V(r)$ is a minimum when the electron is closest to a nucleus (or ion).

In a solid, a valence electron is acted upon by a potential which is characteristic of the whole solid, not just a nearby fixed nucleus. That is, the electron is shared by all the nuclei. The potential function and the electron are said to be **delocalised**. Because the potential function is periodic, the solution to the Schrödinger equation also has periodic form where here we are talking about periodicity of the crystal lattice. This is the **Bloch theorem**. The solution to the time-independent Schrödinger equation (i.e., the eigenfunctions or state function) of the electrons has the three-dimensional vector form:

$$\psi_k(r) = \mu_k(r)e^{ik\bullet r} \quad \text{where} \quad k = \frac{2\pi}{\lambda}$$

When we talk about a valence electron interacting with the regular spacing of nuclei in a crystal lattice, $\psi_k(r)$ has a periodic form characteristic of the lattice. The function e^{ikr} describes the wave-like character of the electron as a matter wave distributed throughout the crystal. The function $\psi_k(r)$ represents the localised modulation of this wave near the relatively fixed ions in the solid. Expressed in terms of the crystal lattice, $\psi_k(r)$ is called the **Bloch function**. This periodicity in $\psi k(r)$ is the reason why we have been so interested in applying **periodic boundary conditions** to our models of solids.

1.3.8 Crystal Potential

In general, at $k = \pm\pi/a$ there is a discontinuity in the energy dispersion as the electrons reach the top of a band. Consider first the simple case of a zero potential. There are no energy bands for the zero potential, but we can still imagine there being bands which touch each other so that a free electron may occupy any level and hence any value of k. This is called the **empty lattice model**.

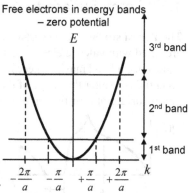

Free electrons in energy bands – zero potential

When the crystal potential is added, we have what is called the **nearly free electron model**. The effect of the crystal potential is to create energy gaps in the dispersion relation.

For free electrons, the wave functions are travelling waves. Because of the periodicity of the crystal, at $k = \pm\pi/a$, wave functions combine by a process called **Bragg reflection** to form standing waves (not travelling waves).

Free electrons in energy bands – with crystal potential

Bragg reflection occurs because of the way an electron wave function interacts with the periodic potential of the positively charged nuclei of the crystal lattice. Whenever an electron wave encounters a nuclei, it is partly reflected (much like

$$E = \frac{\hbar^2}{2m}k^2$$

gap

The region in k space between $k = \pm\pi/a$ is called the first **Brillouin zone**. The region of k space between $k = \pm\pi/a$ is the second Brillouin zone, and so on.

the case of the **step potential** where $E > V_o$). When the wavelength of the electrons is an integral number of the periodicity a of the nuclei, constructive interference for the reflected waves occurs, so that a standing wave pattern is produced. This leads to a departure from the parabolic dispersion curve at $k = \pm n\pi/a$ and the formation of forbidden energy states.

1.3.9 Fermi Surface for Solids

The **Fermi surface** is the surface in k space, at 0K, in which all the states are filled with valence electrons. This may not necessarily be at the top of a band since a band may not necessarily be full. The effect of the crystal potential is to alter the shape of the density of states and the shape of the Fermi surface at near the top of the band.

As the number of valence electrons increases, the shape of the Fermi surface in k space changes from an initial sphere to one which depends upon the crystal geometry of the solid.

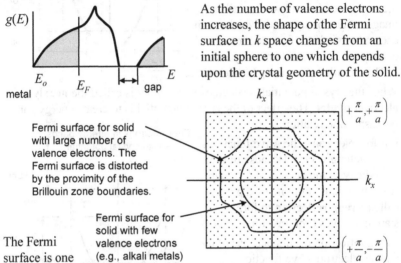

Fermi surface for solid with large number of valence electrons. The Fermi surface is distorted by the proximity of the Brillouin zone boundaries.

Fermi surface for solid with few valence electrons (e.g., alkali metals)

The Fermi surface is one of many surfaces of constant energy, but is distinguished by being that surface which marks the boundary between filled and unfilled states.

Strictly speaking, E_F is the energy for which the occupation number in the **Fermi–Dirac distribution** is ½. In a metal, E_F is usually taken to be the highest filled state because the difference in energy between the highest filled state and the next available state is very small – because these states are located within a band of quasi-continuous states. In an insulator, the next available state is separated from the highest filled state by a relatively large E_g. E_F is the energy halfway in the gap between the highest filled state and the next available state. Actually, E_F for both metals and insulators does not correspond to any physical electron state – it is a value of energy that lies between the highest filled state and the next available state. E_F is a theoretical concept; no electron need actually be present at E_F.

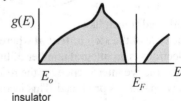

1.3.10 Electrical Conductivity

The velocity of a free electron depends upon the energy level of the state which it occupies. For a free electron, the velocity can be expressed in terms of the momentum p:

$$v = \frac{p}{m}$$

but $p = \hbar k$

thus $v = \frac{\hbar k}{m}$

The significance of this is that v is proportional to k. Hence, the **Fermi surface** can be equally drawn in either **k space** or **velocity space**.

When an electric field E is applied to a free electron in a conductor, the electron experiences a force $-q_e E$ which in terms of the field is written:

$$F = -q_e E$$

$$= \frac{dp}{dt}$$

$$= \hbar \frac{dk}{dt}$$

$$\frac{dk}{dt} = \frac{-q_e E}{\hbar}$$

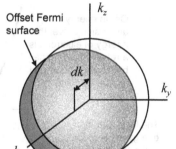

The electron absorbs energy from the field as kinetic energy. The application of the field E produces a dk in time dt where dt can be identified as the **relaxation time** (the time between collisions). Since it is electrons close to the Fermi surface that contribute to electric current, it is not surprising to learn that the conductivity of a metal depends largely upon the shape of the density of states and the position of the Fermi energy.

When a free electron is accelerated by an electric field, it moves to a higher value of k and thus experiences a downward shift in wavelength and an increased velocity v. The presence of the periodic crystal potential only provides a different modulation to the wave function for each value of k. Electrons do not continue to be excited to higher and higher states in the presence of E because of collisions (resistance) brought about by imperfections in the crystal lattice that disrupt the periodicity of the crystal potential and the presence of thermally induced lattice vibrations called **phonons**.

1.3.11 Semiconductors

When most of the states in a band are filled, the remaining vacant states are called **holes**. Holes act like electrons with a positive charge in the sense that they can appear to move and have an effective mass much like an electron.

The defining feature of a **semiconductor** is the size of the energy gap between the valence band and the conduction band. At room temperatures, there is sufficient thermal energy kT to excite some electrons from the valence to the conduction band (E_g in a semiconductor is on the order of 2 eV). The remaining valence band is mostly filled, and, hence, it is of more interest to represent this band in terms of the holes left behind by the electrons that have gone to the conduction band.

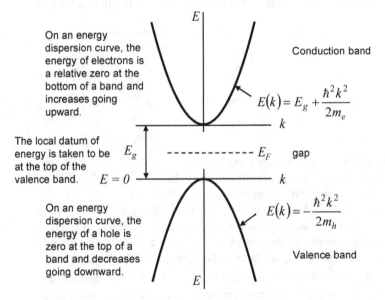

On an energy dispersion curve, the energy of electrons is a relative zero at the bottom of a band and increases going upward.

Conduction band

$$E(k) = E_g + \frac{\hbar^2 k^2}{2m_e}$$

The local datum of energy is taken to be at the top of the valence band. $E = 0$

E_g E_F gap

On an energy dispersion curve, the energy of a hole is zero at the top of a band and decreases going downward.

$$E(k) = -\frac{\hbar^2 k^2}{2m_h}$$

Valence band

In an **intrinsic semiconductor**, at 0K, the material is an insulator and the valence band is full and the conduction band is empty. As the temperature is raised, some electrons are excited into the conduction band, leaving behind holes in the valence band. Both the free electrons in the conduction band and the holes in the valence band are **current carriers** and, in the presence of an electric field, will move and establish an **electric current**.

1.3.12 Intrinsic Semiconductors

For conduction in a semiconductor, it is the tail end of the **Fermi–Dirac distribution** which is of most significance for the existence of free electrons in the conduction band. In the tail, the Fermi-Dirac distribution can be approximated by the **Maxwell–Boltzmann distribution** by neglecting the 1 in the denominator: $f(E) = e^{-(E-E_F)/kT}$

Fermi–Dirac distribution

$$f(E) = \frac{1}{e^{(E-E_F)/kT} + 1}$$

The electrical **conductivity** of a semiconductor is determined by the concentration of mobile or free charge carriers: holes in the valence band and electrons in the conduction band.

The number of electrons in the conduction band from E_C to ∞ can be determined from:

$$n_e = \int_{E_C}^{\infty} f(E)g(E)dE$$

$$= \int_{E_C}^{\infty} e^{-(E-E_F)/kT} \frac{1}{2\pi^2} \left(\frac{2m}{\hbar^2}\right)^{\frac{3}{2}} (E-E_g)^{\frac{1}{2}} dE$$

$$= \frac{1}{2\pi^2} \left(\frac{2m}{\hbar^2}\right)^{\frac{3}{2}} e^{(E_F/kT)} \int_{E_C}^{\infty} (E-E_g)^{\frac{1}{2}} e^{-E/kT} dE$$

$$= 2\left(\frac{2mkT}{2\pi\hbar^2}\right)^{\frac{3}{2}} e^{(E_F/kT)} e^{-E_g/kT}$$

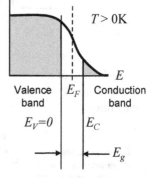

A similar calculation for holes in the valence band, but using Fermi–Dirac statistics, yields the concentration of holes as a function of temperature: $n_p = 2\left(\frac{2m_h kT}{2\pi\hbar^2}\right)^{\frac{3}{2}} e^{-E_F/kT}$

Since the number of holes equals the number of free electrons in the conduction band, a value for E_F can be determined and the addition of n and p gives the concentration of charge carriers in the semiconductor.

$$n = 2\left(\frac{kT}{2\pi\hbar^2}\right)^{\frac{3}{2}} \left(\frac{m_e}{m_h}\right)^{\frac{3}{4}} e^{-E_g/2kT}$$

The important feature of this expression is the exponential rise in n as a function of temperature T (the $T^{3/2}$ factor being small in comparison to the $e^{-E/kT}$ term). The availability of an equal number of holes and electrons as charge carriers is characteristic of an **intrinsic semiconductor**.

1.3.13 P- and N-Type Semiconductors

The addition of certain impurities (doping) to a pure semiconductor can increase conductivity.

1. Introduction of a phosphorous atom (5 valence electrons)

P atom is still electrically neutral.

The now available electron is free to wander around in the **conduction band**.

Silicon lattice

This results in increased conductivity due to availability of negative <u>mobile</u> charge carriers.

and hence is called an

n-type semiconductor.

2. Introduction of a boron atom (3 valence electrons)

B atom is still electrically neutral.

The hole appears to move around as a valence electron falls into it, thus creating a new hole where the electron came from.

This results in increased conductivity due to excess of positive <u>mobile</u> charge carriers

and hence is called a

p-type semiconductor.

In both types of semiconductors, the increased conductivity arises due to the deliberate increase in the number of *mobile* charge carriers – all still electrically neutral material.

• The **majority carriers** in an n-type material are electrons; the majority carriers in a p-type material are holes.

• Both types have thermally generated electrons and holes which are called **minority carriers**.

1.3.14 Extrinsic Semiconductors

When the concentration of holes is equal to the concentration of electrons, the semiconductor is said to be **intrinsic**. This also happens with lightly doped semiconductor materials. For a more heavily doped semiconductor, the concentration of carriers supplied by the impurity atoms is sufficiently large to exceed those produced by thermal effects and the material is either p-type or n-type, depending on the doping material, and is termed **extrinsic**.

Whether a semiconductor is intrinsic or extrinsic depends upon the temperature. At high temperatures, the semiconductor tends to become intrinsic since the number of thermally activated carriers increases (exponentially). At very low temperatures, there may be insufficient thermal energy to keep the normally mobile carriers from the impurity atoms in the conduction band and they remain attached to their parent atoms. In this case, the conductivity is reduced.

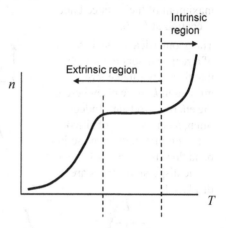

The conductivity of a semiconductor depends upon the motion of the charge carriers in the presence of an electric field. For an n-type material, the electrical conductivity can be expressed in terms of the mean **relaxation time** (time between collisions):

$$\sigma_e = \frac{n q_e^{\,2} \tau}{m_e}$$

The concentration of mobile electrons, n, in a semiconductor is typically on the order 10^{15} cm^{-3} – much smaller than that of a metal, but high enough to provide many interesting and useful phenomena.

Another useful measure of performance of a semiconductor is the **mobility** μ of the charge carriers. This is defined as the drift velocity per unit of field strength. The drift velocity for electrons (n-type) is given by:

$$v_d = \frac{-q_e \tau}{m} E$$

and hence: $\mu = \dfrac{-q_e \tau}{m}$ and $\sigma = -n q_e \mu$

1.3.15 Direct and Indirect Band Gap Semiconductors

The **energy dispersion** curves of a real semiconductor may not necessarily take the parabolic form given by:

This is the simplest form of the energy dispersion where the minimum of the conduction band and the maximum of the valence band is centred at $k = 0$.

$$E(k) = E_g + \frac{\hbar^2 k^2}{2m_e}$$

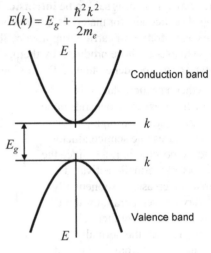

The geometrical crystal structure of a real semiconductor means that there are often energy minima in k space elsewhere in the energy spectrum. Indeed, often, as in the case of silicon, the minimum in the conduction band does not occur at $k = 0$ and so the dispersion curves are displaced:

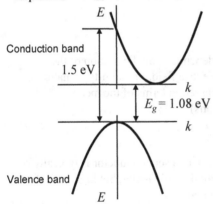

Such a material is called an **indirect band gap semiconductor**. The significance of this is found when a semiconductor absorbs a **photon**. For a direct band gap semiconductor, where the energy minima are centred at $k = 0$, if the incoming photon has an energy greater than the band gap, then an electron in the valence band may be excited into the conduction band in the process of absorption.

Because k is associated with momentum, only vertical transitions from the valence to the conduction band are allowed (to conserve momentum during the absorption process). In an indirect band gap semiconductor, conservation of momentum is achieved by the electron absorbing both a photon and a quantum of mechanical vibration called a **phonon** at the same time. The momentum required to provide the shift in k is provided by the phonon.

1.4 X-Ray Diffraction

Summary

$$eV_K = \frac{hc}{\lambda_{edge}}$$ Absorption edge

$$\mu_m = \frac{N_A}{A}\sigma$$ Mass attenuation coefficient

$$\frac{I_s}{I_o} = \frac{C}{R^2}\left(1 + \cos^2 2\theta\right)$$ X-ray scattering

$$\lambda = 2d\sin\theta$$ Bragg diffraction condition

$$\frac{1}{d} = |\mathbf{S}|$$ Scattering vector

$$f = \int_0^\infty \rho(r)4\pi r^2 \frac{\sin 2\pi s r}{2\pi s r}\, dr$$ Atomic scattering factor

Structure factor (unit cell)

$$F_{hkl} = \frac{E_{uc} \text{ of field scattered by unit cell}}{E \text{ Electric field scattered by electron at origin of unit cell}}$$

$$= \sum f_n e^{2\pi i(hx_n + ky_n + lz_n)}$$

Structure factor (crystal)

$$F_{hkl} = \sum_{n=1}^{N/4} f_n\left[\left(e^{2\pi i(hx_n + ky_n + lz_n)}\right)\left(1 + e^{\pi i(h+k)} + e^{\pi i(h+l)} + e^{\pi i(k+l)}\right)\right]$$

$$B = \frac{2kT}{m}\overline{\frac{1}{f^2}}$$ Debye–Waller factor

1.4.1 X-Rays

X-rays are produced by electrons being decelerated from an initial high speed by collisions with a target material. Electrons may be produced by **thermo ionic emission** and given a high velocity by the application of an electric field.

When an electron q_e collides with a target material, it is rapidly decelerated and a photon is emitted. The wavelengths of the photons are mainly in the x-ray region of the electromagnetic spectrum. The most rapid decelerations result in the shortest wavelength photons. For other collisions, the electron may lose energy via the emission of photons of longer wavelength, and may also lose energy to heat by increasing the vibrational internal energy of the target. The result is a **continuous spectrum** of photon energies with a minimum wavelength dependent upon the kinetic energy of the electrons.

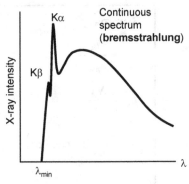

The high energy limit is determined by the high voltage (kV) of the electron source.

$$eV = h\upsilon_{max} = \frac{hc}{\lambda_{min}}$$

$$\lambda_{min} = \frac{hc}{eV} = \frac{12.4 \times 10^3}{V} \quad \text{A}$$
kV

Incoming electrons may also ionise the atoms of the target by ejecting bound electrons from within material. Some of these ejected electrons may come from the innermost energy levels which, in solid, can have energies in excess of 100,000 eV. An outer electron can fall into this vacancy and emit a photon in the process.

X-rays resulting from filling of K shell vacancies by an electron from the L shell are called Kα x-rays. X-rays from M to K shell transitions are Kβ, and those from N (and higher) to K transitions are Kγ. Similarly, transitions from M to L are Lα, N (and higher) are Lβ. These emissions result in sharp peaks in the overall energy spectrum of emission which are collectively called **characteristic radiation**.

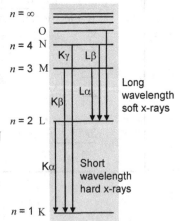

1.4.2 X-Ray Intensity

The intensity of the **continuous radiation** is determined by the voltage
used to accelerate the incident electrons.

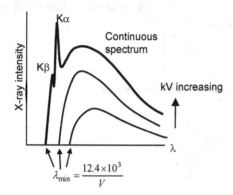

$$\lambda_{min} = \frac{12.4 \times 10^3}{V}$$

The intensity of the **characteristic radiation** or (**x-ray line spectra**) emission
from a target material is determined by the relative number of states from
which transitions can occur.

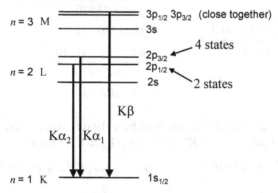

The relative intensities of $K\alpha_1$ to
$K\alpha_2$ are found by considering the
number of states:

$$\frac{I(\alpha_1)}{I(\alpha_2)} = \frac{4}{2} = 2$$

The relative intensities of $K\alpha$ to $K\beta$
cannot be found this way. What can
be said is that the probability of a
change in quantum number n of 2 is
less than a change in n of 1, and so:

$$I(K\alpha) > I(K\beta)$$
$$\frac{I(K\alpha)}{I(K\beta)} = 9.4 \ (Cu)$$

1.4.3 Absorption Edge

An incoming electron incident on a target may eject an electron from a K shell in the target material. K radiation can only occur when the potential used to accelerate the electron is above a critical value which depends upon the ionisation potential of the K shell of the atom.

$$eV_K = \frac{hc}{\lambda_{edge}}$$

For example, the K and L **absorption edges** for barium are:

$$\lambda_{edge}$$

K	0.3310 Å
L_I	2.0680 Å
L_{II}	2.2048 Å
L_{III}	2.3629 Å

The energies associated with the electron shells are found from:

$$E_K = \frac{hc}{\lambda_{edge}}$$

$$= \frac{6.63 \times 10^{-34}\left(3 \times 10^8\right)}{0.3310 \times 10^{-10}}$$

$$= 6 \times 10^{-15}\,J$$

$$E_{LI} = \frac{6.63 \times 10^{-34}\left(3 \times 10^8\right)}{2.2048 \times 10^{-10}}$$

$$= 9.021 \times 10^{-16}\,J$$

$$E_{LIII} = \frac{6.63 \times 10^{-34}\left(3 \times 10^8\right)}{2.3629 \times 10^{-10}}$$

$$= 8.42 \times 10^{-16}\,J$$

The wavelengths emitted are thus determined from the difference in the energies and so for $BaK\alpha_1$ (L_{III} to K) and $BaK\alpha_2$ (L_{II} to K) radiations:

$$\Delta E_{K\alpha_1} = 6 \times 10^{-15} - 9.021 \times 10^{-16}$$

$$= 5.1 \times 10^{-15}\,J$$

$$= \frac{hc}{\lambda_1}$$

$$\lambda_1 = \frac{6.63 \times 10^{-34}\left(3 \times 10^8\right)}{5.1 \times 10^{-15}}$$

$$= 3.9 \times 10^{-11}\,m$$

$$= 0.390\mathring{A}\ K\alpha_1$$

$$\Delta E_{K\alpha_2} = 6 \times 10^{-15} - 8.42 \times 10^{-16}$$

$$= \frac{hc}{\lambda_2}$$

$$\lambda_2 = 0.386\mathring{A}\ K\alpha_1$$

1.4.4 Fluorescence Yield

The x-rays associated with absorption edges are **characteristic radiation** of the target material as distinct from the continuum of wavelengths that arise from decelerations of the incoming electrons.

Characteristic radiation is very important because it is a source of radiation of nearly single wavelength that can be used for diffraction experiments and can also be used to characterise the target material.

The most efficient operating voltage V of an x-ray tube occurs when the ratio of the intensity of the characteristic radiation to that of the continuous radiation is a maximum. This ratio is given by:

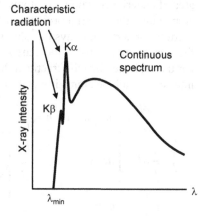

$$\frac{I_{characteristic}}{I_{continuous}} = K\frac{\left(V-V_K\right)^{1.7}}{V^2}$$

$$V = 6.5V_K \text{ at maximum}$$

When an electron is ejected, the resulting vacancy can be filled by a transition of an electron from an L or M shell, resulting in Kα or Kβ radiation, or from higher shells resulting in longer wavelength x-rays.

A transition from L to K may also occur without the emission of an x-ray photon. In this case, another electron, called an **Auger electron**, is emitted from a higher level state to maintain the energy balance.

The K fluorescence yield is the fraction of vacancies in the K shell that actually result in x-ray emission. Similarly, the L fluorescence yield is the fraction of vacancies in the L shell that result in actual x-ray emission. Atoms with a low atomic number (<30) have a low fluorescence yield and the Auger process dominates over characteristic x-ray emission or x-ray fluorescence (**XRF**).

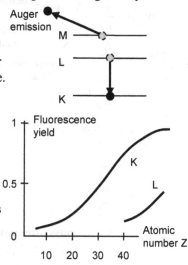

1.4.5 X-Ray Absorption

When x-rays pass through a material, the intensity of the beam is reduced by absorption and scattering.

Consider a beam of x-rays incident on a unit area of material containing N atoms. Each atom within the material presents a cross-sectional area σ to the rays through which the x-rays cannot pass. If there are n atoms per unit volume, then the fractional loss of intensity for all the atoms is equal to:

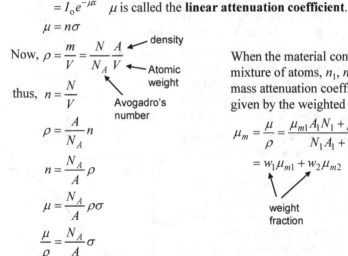

$$-\frac{\delta I}{I} = n\sigma\delta x \text{ where } n = \frac{N}{V}$$

$$\int_{I_o}^{I_x} -\frac{1}{I}dI = \int_0^x n\sigma dx$$

$$\ln\frac{I_x}{I_o} = -n\sigma x$$

$$I_x = I_o e^{-n\sigma x}$$

$$= I_o e^{-\mu x} \quad \mu \text{ is called the } \textbf{linear attenuation coefficient.}$$

$$\mu = n\sigma$$

Now, $\rho = \dfrac{m}{V} = \dfrac{N}{N_A}\dfrac{A}{V}$ ← density ← Atomic weight ← Avogadro's number

thus, $n = \dfrac{N}{V}$

$$\rho = \frac{A}{N_A}n$$

$$n = \frac{N_A}{A}\rho$$

$$\mu = \frac{N_A}{A}\rho\sigma$$

$$\frac{\mu}{\rho} = \frac{N_A}{A}\sigma$$

$$= \mu_m$$

When the material contains a mixture of atoms, $n_1, n_2, ...,$ the mass attenuation coefficient is given by the weighted sum:

$$\mu_m = \frac{\mu}{\rho} = \frac{\mu_{m1}A_1N_1 + \mu_{m2}A_2N_2}{N_1A_1 + N_2A_2}$$

$$= w_1\mu_{m1} + w_2\mu_{m2}$$

weight fraction

$= \mu_m$ **mass attenuation coefficient.** μ_m is independent of the state (liquid, gas, etc.) of matter.

1.4.6 Attenuation of X-Rays

Attenuation of x-rays by a target material occurs due to scattering and photoelectric effects.

In **coherent scattering**, an x-ray sets an electron in a target atom into oscillation and x-rays at the same wavelength are re-radiated in all directions. This is called **Thomson scattering**.

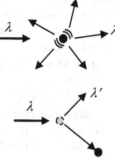

In **incoherent scattering**, energy from the x-ray is transferred to an electron (but not enough to cause a transition as in photoelectric attenuation), and a longer wavelength x-ray is re-radiated. This is called the **Compton effect**.

In the Compton effect, an x-ray photon of energy $h\nu$ collides with a *free* electron in the target material, thus transferring some of its energy to the initially stationery electron, which gains kinetic energy. The scattered x-ray photon is left with less energy and, hence, a lower frequency and longer wavelength. The amount of energy loss depends upon the angle of collision and so the Compton shift in wavelength depends upon the angle of scatter.

At long wavelength incident radiation, Thomson scattering is the dominant process. At shorter wavelengths, Compton scattering dominates. At x-ray wavelengths, neither effect is significant compared to attenuation by photoelectric absorption.

In the photoelectric effect, an incoming x-ray photon has sufficient energy to eject a bound electron from an energy shell. The electron is given kinetic energy which is transmitted to the lattice. The filling of the resulting vacancy is accompanied by emission of x-rays or **Auger electrons**.

Photoelectric attenuation of the incident ray is a maximum when the x-ray photons have an energy closest to the ionisation energy of an electron shell. That is, $\lambda > \lambda_{edge}$.

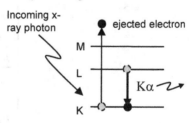

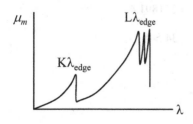

1.4.7 Example

A 12 μm foil is placed in a beam of x-rays from a Co target. The measured count rates of the CoKα and CoKβ radiations transmitted through the foil are 2500 and 32 counts per second. Calculate the relative intensity of the CoKα and CoKβ radiations in the incident beam given the following data:

$A_{Fe} = 55.85 \times 10^{-3}$ kg/mol
$N_A = 6.022 \times 10^{23}$ /mol
$\rho_{Fe} = 7870$ kg/m³
Attenuation cross section Fe at:
CoKα $= 0.5216 \times 10^{-24}$ m²
CoKβ $= 3.2040 \times 10^{-24}$ m²

Answer:

$$I = I_o e^{-\mu_m \rho t}$$

$$\frac{2500}{I_{o\alpha}} = e^{-5.62486(7870)12\times10^{-6}}$$

$$= e^{-0.531}$$

$$\frac{32}{I_{o\alpha}} = e^{-34.548(7870)12\times10^{-6}}$$

$$= e^{-3.263}$$

$$\frac{I_{o\alpha}}{I_{o\beta}} = \frac{2500e^{-3.263}}{32e^{-0.531}}$$

$$\approx 5$$

Attenuation coefficients:

$$\sigma = 0.5216 \times 10^{-24} \text{ /unit cell}$$

$$\mu = \sigma N \longrightarrow \text{No. atoms per unit cell}$$

$$\rho = \frac{NA}{N_A}$$

$$7870 = \frac{55.85 \times 10^{-3} N}{6.022 \times 10^{23}}$$

$$N = 8.486 \times 10^{28}$$

$$\mu_{K\alpha} = 0.5216 \times 10^{-24}\left(8.486 \times 10^{28}\right)$$

$$= 44262.9$$

$$\frac{\mu_{K\alpha}}{\rho} = 5.624$$

$$\frac{\mu_{K\beta}}{\rho} = 3.204 \times 10^{-24}\left(8.486 \times 10^{28}\right)$$

$$= 271891.4$$

$$\frac{\mu_{K\alpha}}{\rho} = 34.548$$

1.4.8 X-Ray Scattering – Electron

Consider the case of scattering of x-rays by a single isolated electron. The incoming radiation is represented by transverse electric field vectors of amplitude E_{oY} and E_{oz}.

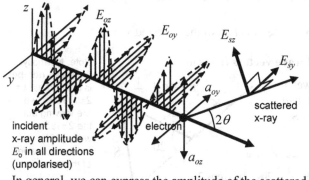

E_{oz} makes the electron vibrate with an acceleration a_{oz} in the z direction. E_{oy} makes the electron vibrate with an acceleration a_{oy} in the y direction. Scattered field has amplitude E_{sz} and E_{sy}.

In general, we can express the amplitude of the scattered ray E_S in terms of the distance R from the electron location.

$$E_S = \frac{-q_e a}{4\pi\varepsilon_o c^2 R} \quad \text{where } a \text{ is the acceleration parallel to } E_S$$

The amplitude of the scattered waves in z can be expressed in terms of the force acting on the electron by the field E_{oz}:

$$ma_{oz} = -q_e E_{oz} \text{ force acting on electron from field } E_{oz}$$

$$a_{sz} = a_{oz}\cos 2\theta \quad \text{acceleration } \parallel \text{ to } E_{sz}$$

amplitude $$E_{sz} = \frac{-q_e}{4\pi\varepsilon_o c^2 R} a_{oz}\cos 2\theta = \frac{-q_e^2 E_{oz}}{4\pi\varepsilon_o c^2 Rm}\cos 2\theta$$

similarly $$E_{sy} = \frac{-q_e^2 E_{oy}}{4\pi\varepsilon_o c^2 Rm}$$

Note, in this example, $a_{oy} = a_{sy}$.

Note the inverse relationship with m. Scattering from an electron is much greater than scattering from a proton.

Intensity I (Wm^{-2}) is proportional to E^2: $$\overline{I} = \varepsilon_o c \frac{1}{2}\left(E_z^2 + E_y^2\right)$$

$$\overline{I_s} = \frac{1}{2}c\varepsilon_o\left(\overline{E_{sz}^2} + \overline{E_{sy}^2}\right)$$

$$\frac{I_s}{I_o} = \frac{C}{R^2}\left(1 + \cos^2 2\theta\right) \quad \text{where } C \text{ is all the constants}$$

I_s/I_o

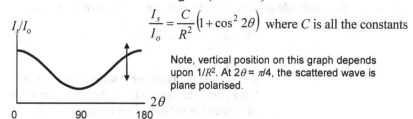

Note, vertical position on this graph depends upon $1/R^2$. At $2\theta = \pi/4$, the scattered wave is plane polarised.

2θ

0   90   180

1.4.9 Wave Vectors

Consider two electrons spaced a distance d apart. X-rays are incident on both and are scattered.

Let $|\mathbf{K_o}| = \dfrac{1}{\lambda}$ and be in the direction of the incident x-rays

and $|\mathbf{K_s}| = \dfrac{1}{\lambda}$ and be in the direction of the scattered x-rays.

Both K's are called **wave vectors** (units m^{-1}).

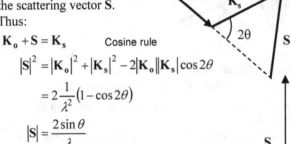

The difference between the two vectors is called the scattering vector **S**. Thus:

$$\mathbf{K_o} + \mathbf{S} = \mathbf{K_s} \qquad \text{Cosine rule}$$

$$|\mathbf{S}|^2 = |\mathbf{K_o}|^2 + |\mathbf{K_s}|^2 - 2|\mathbf{K_o}||\mathbf{K_s}|\cos 2\theta$$

$$= 2\frac{1}{\lambda^2}(1 - \cos 2\theta)$$

$$|\mathbf{S}| = \frac{2\sin\theta}{\lambda}$$

Note: Normally, the **wave number** is defined as: $k = 2\pi/\lambda$. Here, we are leaving out the 2π factor and writing K to remind ourselves of this omission.

$$|\mathbf{K_s}| = |\mathbf{K_o}| = \frac{1}{\lambda}$$

At the **Bragg condition** for constructive interference,

$$\left.\begin{array}{l} \lambda = 2d\sin\theta \\ \text{thus} \quad \dfrac{1}{d} = \dfrac{2\sin\theta}{\lambda} \\ \qquad = |\mathbf{S}| \end{array}\right\}$$

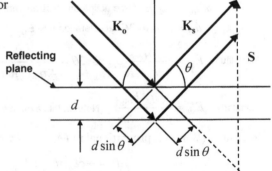

Reflecting plane

We define the amplitude of the vector **d*** as being the reciprocal of d, so that at the Bragg condition,

$$\mathbf{S} = \mathbf{d}* \qquad \text{where } |\mathbf{d}*| = \frac{1}{d}$$

d* is called the **reciprocal lattice vector** and has a particular convenience during the analysis of diffraction patterns from crystals.

1.4.10 X-Ray Scattering – Atom

Let an atom be represented by a spherically symmetric distribution of charge (electron cloud) surrounding the nucleus. Consider the wave vectors $\mathbf{K}$ of an incident beam of x-rays and scattered x-rays from a small volume of charge (i.e., not necessarily a single electron) at some distance $\mathbf{r}$ from the centre of the atom and an electron at the centre of the atom (if one could exist there).

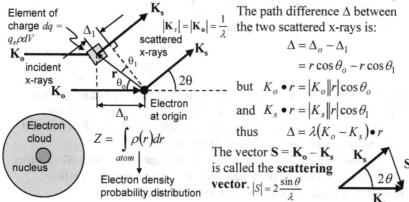

Element of charge $dq = q_e \rho dV$

$\mathbf{K_o}$ incident x-rays

$|\mathbf{K}_s| = |\mathbf{K}_o| = \dfrac{1}{\lambda}$ scattered x-rays

$Z = \displaystyle\int_{atom} \rho(r)\,dr$

Electron density probability distribution

The path difference Δ between the two scattered x-rays is:

$$\Delta = \Delta_o - \Delta_1$$
$$= r\cos\theta_o - r\cos\theta_1$$

but $\quad K_o \bullet r = |K_o||r|\cos\theta_o$

and $\quad K_s \bullet r = |K_s||r|\cos\theta_1$

thus $\qquad \Delta = \lambda(K_o - K_s)\bullet r$

The vector $\mathbf{S} = \mathbf{K_o} - \mathbf{K_s}$ is called the **scattering vector.** $|S| = 2\dfrac{\sin\theta}{\lambda}$

A wave of amplitude E_o can be expressed as: $E = E_o e^{i(2\pi Kx + \phi)}$

The phase difference between two waves is $\dfrac{2\pi}{\lambda}\Delta = 2\pi(\mathbf{S}\bullet\mathbf{r})$

The ratio of the charge on the electron and at the element dV is: $\dfrac{dq}{-q_e} = \rho dV$

Thus the ratio of the amplitudes for the two scattered waves is:

$$\dfrac{E_s}{E_e} = \dfrac{\rho dV e^{i2\pi(Kx+\mathbf{S}\bullet\mathbf{r})}}{e^{i2\pi Kx}}$$

$$= \rho dV e^{i2\pi(\mathbf{S}\bullet\mathbf{r})}$$

$$f = \int_{atom} \rho e^{i2\pi(\mathbf{S}\bullet\mathbf{r})}\,dV$$

To perform this integral, the sum of the volume elements with the same value of $\mathbf{S}\bullet\mathbf{r}$ is found and summed over all angles ϕ.

$\mathbf{S}\bullet\mathbf{r} = |S||r|\cos\phi$

$$f_{atom} = \int_{r=0}^{\infty}\int_{\phi=0}^{\pi} \rho(r)e^{2\pi iSr\cos\phi}\,2\pi r^2 \sin\phi\,d\phi d\theta$$

The **atomic scattering factor** f is the ratio between the amplitude E_s of the scattered x-ray by an atom (consisting of many electrons) to that of E_e of a single electron if it were located at the centre of the atom.

$$f = \int_0^{\infty} \rho(r)4\pi r^2 \dfrac{\sin 2\pi sr}{2\pi sr}\,dr$$

Note: At $S = 0$, $2\theta = 0$, forward scattering and $f_{atom} = Z$.

1.4.11 Reciprocal Lattice

Consider a plane of atoms within the unit cell. Any plane is represented by the **Miller indices** (*hkl*). The spacing between planes is given by:

$$d = \frac{1}{\sqrt{\left(\frac{h}{a}\right)^2 + \left(\frac{k}{b}\right)^2 + \left(\frac{l}{c}\right)^2}}$$

The reciprocal lattice vector is defined such that: $|d*| = \dfrac{1}{d}$

$$\text{Thus: } |d*|^2 = \frac{h^2}{a^2} + \frac{k^2}{b^2} + \frac{l^2}{c^2}$$

The allowed values of $|d*|$ are determined by the planes that exist within the crystal structure and their spacing. Only certain values of $|d*|$ are possible. For example, for the (100) plane, $|d*|^2 = 1/a^2$. The direction of $d*$ is in the same direction as **S** at the **Bragg condition**.

The allowed values of $|d*|$ can be expressed graphically, much like a lattice of points where each point in the lattice, called the **reciprocal lattice**, represents a plane in the real crystal **space lattice**.

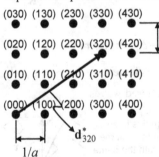

The reciprocal lattice is a lattice of values of **S** corresponding to possible diffraction (Bragg) conditions. A distance *d* between the origin and a point in the reciprocal lattice represents a possible value of $|S|$ where: $|S| = \dfrac{1}{d}$

The direction of $d*$ is perpendicular to the *hkl* plane at the Bragg condition.

A point at distance $d*$ from (000) in the reciprocal lattice is thus represented as a vector: $d^*_{hkl} = h\mathbf{a}^* + k\mathbf{b}^* + l\mathbf{c}^*$

More generally:

$|\mathbf{a}*| = |d^*_{100}|$ $\mathbf{a}*$ is $\perp$ to the 100 plane

$|\mathbf{b}*| = |d^*_{010}|$ $\mathbf{b}*$ is $\perp$ to the 010 plane

$|\mathbf{c}*| = |d^*_{001}|$ $\mathbf{c}*$ is $\perp$ to the 001 plane

For an orthogonal unit cell:

$$|\mathbf{a}*| = \frac{1}{a}; |\mathbf{b}*| = \frac{1}{b}; |\mathbf{c}*| = \frac{1}{c}$$

$$\mathbf{a}^* = \frac{\mathbf{b} \times \mathbf{c}}{(\mathbf{a} \times \mathbf{b}) \bullet \mathbf{c}}$$

$$\mathbf{b}^* = \frac{\mathbf{c} \times \mathbf{a}}{(\mathbf{a} \times \mathbf{b}) \bullet \mathbf{c}}$$

$$\mathbf{c}^* = \frac{\mathbf{a} \times \mathbf{b}}{(\mathbf{a} \times \mathbf{b}) \bullet \mathbf{c}}$$

The denominator is the volume of the unit cell in real space.

1.4.12 Monoclinic Lattice

Calculation of the d spacing for a monoclinic lattice provides an excellent example of the general procedure.

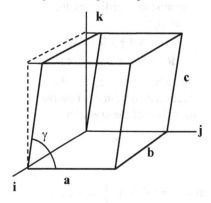

For this structure, one cannot use

$$|a^*| = \frac{1}{a},$$

etc., because it is a non-orthogonal unit cell.

i, j and k are unit vectors. The lattice parameters are thus expressed:

$$b = bi$$

$$a = aj$$

$$c = c\sin\beta\, k + c\cos\beta\, j$$

Cell volume: $V = abc\sin\beta$

First, determine expressions for a^*, b^* and c^*:

$$a^* = \frac{b \times c}{V}$$

$$b \times c = \begin{vmatrix} i & j & k \\ b & 0 & 0 \\ 0 & c\cos\beta & c\sin\beta \end{vmatrix}$$

$$= -j(bc\sin\beta) + i(0) + k(bc\cos\beta)$$

$$= -bc\sin\beta\, j + bc\cos\beta\, k$$

$$a^* = \frac{-bc\sin\beta\, j + bc\cos\beta\, k}{abc\sin\beta}$$

$$= -\frac{1}{a}j + \frac{\cos\beta}{a\sin\beta}k$$

Similarly:

$$b^* = \frac{1}{b}i$$

$$c^* = -\frac{1}{c\sin\beta}k$$

Then determine d^* and hence $|d^*|$ where:

$$d^* = ha^* + kb^* + lc^*$$

$$= -\frac{h}{a}j + \frac{h\cos\beta}{a\sin\beta}k + \frac{k}{b}i + \frac{-l}{c\sin\beta}k$$

$$= \frac{k}{b}i + \frac{-h}{a}j + \frac{(ch\cos\beta - la)}{ac\sin\beta}k$$

$$|d^*| = \frac{1}{d} = d^* \bullet d^*$$

$$= \left(\frac{k^2}{b^2} + \frac{-h^2}{a^2} + \frac{(ch\cos\beta - la)^2}{(ac\sin\beta)^2} \right)^{\frac{1}{2}}$$

$$= \frac{k^2}{b^2} + \frac{l^2}{(c\sin\beta)^2} - \frac{2hl\cos\beta}{ac\sin^2\beta}$$

$$+ \frac{h^2}{(a\sin\beta)^2}$$

1.4.13 Structure Factor

Let the position vectors of the atoms in the unit cell be $\mathbf{R}_1, \mathbf{R}_2, \dots \mathbf{R}_n$ inside the unit cell.

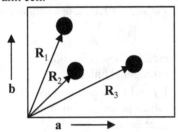

The atomic coordinates are given by:

$$\mathbf{R}_1 = x_1\mathbf{a} + y_1\mathbf{b} + z_1\mathbf{c}$$
$$\mathbf{R}_2 = x_2\mathbf{a} + y_2\mathbf{b} + z_2\mathbf{c},$$

etc., where (x_i, y_i, z_i) are the fractional coordinates relative to the origin of the crystal axes.

Now, $\mathbf{R} = x\mathbf{a} + y\mathbf{b} + z\mathbf{c}$

and $\mathbf{d}^*_{hkl} = h\mathbf{a}^* + k\mathbf{b}^* + l\mathbf{c}^*$

thus $\mathbf{d}^*_{hkl} \bullet \mathbf{R} = hx + ky + lz$ since

$$\left.\begin{array}{cc} \mathbf{b}*\bullet\mathbf{b}=1 & \mathbf{a}*\bullet\mathbf{b}=0 \\ \mathbf{c}*\bullet\mathbf{c}=1 & \mathbf{b}*\bullet\mathbf{a}=0 \\ \mathbf{a}*\bullet\mathbf{a}=1 & \mathbf{c}*\bullet\mathbf{a}=0 \end{array}\right\}$$

Applies to orthogonal and non-orthogonal axes

At the Bragg condition, $\mathbf{S} = \mathbf{d}*$, so the **atomic scattering factor** is:

$$f = \int\limits_{atom} \rho e^{i2\pi(\mathbf{S}\bullet\mathbf{R})}dV = \int\limits_{atom} \rho e^{i2\pi(\mathbf{d}*\bullet\mathbf{R})}dV = \int\limits_{atom} \rho e^{i2\pi(hx+ky+lz)}dV$$

The amplitude of the scattered wave from a unit cell E_{uc} is the sum of the scattered radiation from each atom in the unit cell.

$$E_{uc} = \sum_{uc} E_{atom}$$

atomic scattering factor

$$= E_e \underbrace{\sum_{n} f_n e^{2\pi i(\mathbf{d}^*_{hkl}\bullet\mathbf{R}_i)}}$$

This is the **structure factor** F_{hkl} of the unit cell.

$$F_{hkl} = \frac{E_{uc} \text{ of field scattered by unit cell}}{E \text{ Electric field scattered by electron at origin of unit cell}}$$

$$= \sum f_n e^{2\pi i(hx_n+ky_n+lz_n)}$$

The **structure factor** is measured and interpreted to determine crystal structure in x-ray diffraction experiments. It provides information about the intensity of diffracted (scattered waves at the Bragg condition) x-rays from the crystal lattice.

1.4.14 X-Ray Scattering – Crystal

The intensity of the x-rays scattered by a crystal is proportional to the square of the electric field vector amplitude, which in turn is governed by the structure factor F_{hkl} for the plane under consideration within each unit cell.

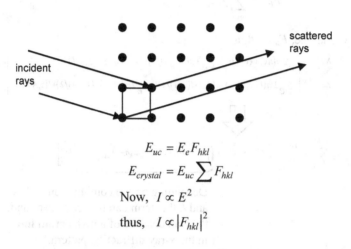

$$E_{uc} = E_e F_{hkl}$$

$$E_{crystal} = E_{uc} \sum F_{hkl}$$

Now, $I \propto E^2$

thus, $I \propto |F_{hkl}|^2$

At the Bragg diffraction condition for interference maxima,

$$I_{max} \propto I_e N^2 |F_{hkl}|^2$$

In an **x-ray diffractometer**, incident x-rays are scattered by the crystal being measured and the angle of incidence (or the wavelength of the incident rays) is slowly varied. The resulting peaks in the intensity of the scattered rays is recorded against the angle 2θ.

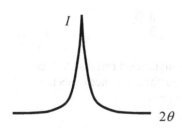

There may be many such peaks in a diffraction pattern, some of them overlapping. The area under a peak is called the **integrated intensity** J.

$$J = \int I(2\theta) d\theta$$

$$\propto N |F_{hkl}|^2$$

The diffraction angles, together with the integrated intensities, allow a diffraction pattern to be indexed and the composition (such as atom type and spacing) of the specimen determined.

1.4.15 X-Ray Intensities

The structure factor for a crystal is calculated by summing the structure factors for each component atom in the unit cell.

$$F_{hkl} = \sum_{n=1}^{N} f_n e^{2\pi i(hx_n + ky_n + lz_n)}$$

For an FCC lattice, F_{hkl} is thus:

$$F_{hkl} = \sum_{n=1}^{N/4} f_n \left[\begin{array}{l} e^{2\pi i(hx_n + ky_n + lz_n)} + e^{2\pi i(h(x_n + 1/2) + k(y_n + 1/2) + l(z_n + 1/2))} \\ + e^{2\pi i(h(x_n + 1/2) + k(y_n) + l(z_n + 1/2))} + e^{2\pi i(h(x_n) + k(y_n + 1/2) + l(z_n + 1/2))} \end{array} \right]$$

$$\Downarrow$$

$$F_{hkl} = \sum_{n=1}^{N/4} f_n \left[\left(e^{2\pi i(hx_n + ky_n + lz_n)} \right) \underbrace{\left(1 + e^{\pi i(h+k)} + e^{\pi i(h+l)} + e^{\pi i(k+l)} \right)}_{} \right]$$

> Depending on the combination of h, k and l, this term can reduce to zero and, thus, an absence of a **diffraction line** in the x-ray diffraction pattern.

Suppose hkl are not all odd or all even: e.g., 100, 110, 211....

$$\begin{array}{l} e^{odd\,\pi i} = -1 \\ e^{even\,\pi i} = +1 \end{array} \quad \text{and} \quad \left(1 + e^{\pi i(h+k)} + e^{\pi i(h+l)} + e^{\pi i(k+l)} \right) = 0$$

Suppose hkl are all odd or all even: e.g., 111, 000, 222.... Then $h + l =$ even, $h + k =$ even, and $k + l =$ even,

$$F_{hkl} = \sum_{n=1}^{N/4} 4 f_n \left[\left(e^{2\pi i(hx_n + ky_n + lz_n)} \right) \right]$$

Recognizing which combinations of hkl are significant can lead to great simplifications in determining the structure factor and hence expected intensities (or even existence) of diffraction lines.

1.4.16 Example

Consider the scattering of x-rays by a metal **BCC lattice**. In this first example, we have similar atoms at (000) and (½ ½ ½).

$$F_{hkl} = fe^{2\pi i(0)} + fe^{2\pi i\left(\frac{h}{2}+\frac{k}{2}+\frac{l}{2}\right)}$$
$$= f\left[1 + e^{\pi i(h+k+l)}\right]$$

when $h + k + l$ is an even number, $F_{hkl} = 2f_{atom}$

when $h + k + l$ is an odd number, $F_{hkl} = 0$

Next consider caesium chloride, where we have a BCC lattice with Cs in the centre (½ ½ ½) and Cl at the corners (000).

$$F_{hkl} = fe^{2\pi i(0)} + fe^{2\pi i\left(\frac{h}{2}+\frac{k}{2}+\frac{l}{2}\right)}$$
$$= f_{Cl} + f_{Cs}e^{\pi i(h+k+l)}$$

when $h + k + l$ is an even number, $F = F_{Cl} + F_{Cs}$ ↙ stronger lines

when $h + k + l$ is an odd number, $F = F_{Cl} - F_{Cs}$ ↖ weaker lines

The intensity of the diffraction peaks is proportional to the square of the structure factors: $I \propto \left(F_{Cl} - F_{Cs}\right)^2$

The diffraction pattern for CsCl is thus a series of alternating strong and weak peaks, or **diffraction lines**.

1.4.17 Example

Consider the scattering of x-rays by a metal with an **FCC lattice** (e.g., Al). In this first example, we have similar atoms at (000), (½ 0 ½), (½ ½ 0) and (0 ½ ½).

$$F_{hkl} = fe^{2\pi i(0)} + fe^{2\pi i\left(\frac{h}{2}+\frac{k}{2}\right)} + fe^{2\pi i\left(\frac{h}{2}+\frac{l}{2}\right)} + fe^{2\pi i\left(\frac{k}{2}+\frac{l}{2}\right)}$$

$$= f\left[1 + e^{\pi i(h+k)} + e^{\pi i(h+l)} + e^{\pi i(k+l)}\right]$$

$$= 4f \quad \text{when } hkl \text{ are all odd or all even}$$

$$= 0 \quad \text{when } hkl \text{ are neither all odd nor all even (i.e., mixed)}$$

In general, we can view structure factors as being orthogonal vectors or complex numbers such that:

$$F_{hkl} = A_{hkl} + jB_{hkl}$$

$$I_{hkl} \propto |F|^2$$

$$\propto \left(A_{hkl}^2 + B_{hkl}^2\right)$$

1.4.18 X-Ray Diffraction – Peak Position

The results of an x-ray diffraction measurement take the form of an x-ray diffraction pattern. This pattern consists of a series of counts from the x-ray intensity sensor (which provides a measure of intensity of the diffracted x-ray beam) as a function of the incident angle expressed as 2θ.

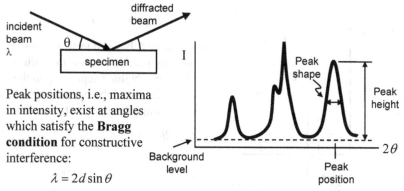

Peak positions, i.e., maxima in intensity, exist at angles which satisfy the **Bragg condition** for constructive interference:

$$\lambda = 2d\sin\theta$$

Peak positions enable the spacing d between crystallographic planes to be measured, as well as **lattice parameters** (size of the unit cell). Changes in lattice parameter as a result of mechanical load can also be measured, leading to a measurement of the **strain** within the specimen.

The peak position for strong sharp peaks is more easily measured than are broad low intensity peaks. Complex fitting procedures may be required to establish the position of a peak, or a series of overlapping peaks.

The precision at which a peak may be measured can be affected by wavelength **dispersion** within the incident beam. A typical source of x-rays is the CuKα wavelength – which consists of two closely spaced frequencies: Kα_1 = 1.5406 Å and Kα_2 = 1.5443 Å. The intensity of Kα_1 is approximately twice that of Kα_2. For a particular value of d, there are thus two closely spaced peaks for each of these wavelength components.

$$\lambda_1 = 2d\sin\theta_1$$
$$\lambda_2 = 2d\sin\theta_2$$
$$\lambda_2 - \lambda_1 = \Delta\lambda = 2d(\sin\theta_2 - \sin\theta_1)$$
$$\approx d\Delta2\theta\cos\theta_{av}$$
$$\Delta2\theta = 2\frac{\Delta\lambda}{\lambda_{av}}\tan\theta_{av}$$

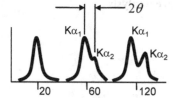

The separation of the Kα peaks for the same value of d becomes larger at high angles.

1.4.19 X-Ray Diffraction – Peak Intensity

The x-ray detector in an x-ray diffractometer is typically a scintillation counter from which accumulated counts in a fixed time period are an indication of intensity of the diffracted beam. The detector (or the sample) is moved in steps of 2θ and the counts accumulated at each step position.

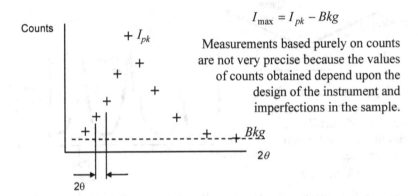

$$I_{max} = I_{pk} - Bkg$$

Measurements based purely on counts are not very precise because the values of counts obtained depend upon the design of the instrument and imperfections in the sample.

A more common measure of the strength of a diffraction line is the **integrated intensity**.

$$J = \int I(2\theta)d2\theta$$
$$= \sum (I(2\theta) - Bkg(2\theta))\Delta 2\theta$$

The integrated intensity is less susceptible to sample imperfections and instrument design.

The intensity of the diffraction lines in the diffraction pattern allows identification of the crystal structure. Absences of peaks in certain positions indicate the presence of various planes within the structure as do the relative intensities of the peaks. This procedure is called **crystal structure analysis**.

1.4.20 X-Ray Diffraction – Peak Shape

Peaks in a diffraction pattern are not infinitely narrow. The spread or width of the peak is determined by the dispersion in the incident radiation and the design of the diffractometer.

A useful measure of the width of a diffraction peak is the width as measured at half the maximum intensity (**HHW**).

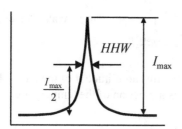

The width at half maximum may be difficult to interpret, especially when there are overlapping peaks resulting from wavelength dispersion in the beam.

Another useful indicator of shape is the integrated breadth B. This is the width of a rectangular peak of the same height and integrated intensity as the real peak.

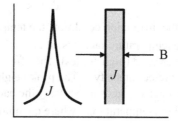

Line broadening can occur due to conditions in the sample, such as the presence of very small crystallites, inhomogeneous strain due to residual stresses, and defects in the crystal structure.

1.4.21 Temperature Effects

Atoms in a unit cell are in constant motion due to their internal energy of vibration resulting from temperature. Atoms will thus oscillate about lattice points and this will affect the Bragg condition for x-ray diffraction.

When temperature is taken into account, a **temperature factor** is applied to the expression for the integrated intensity so that:

$$J = \int I(2\theta)d\theta$$

— **Debye–Waller factor**

$$= K|F_{hkl}|^2 e^{-2B\frac{\sin^2\theta}{\lambda^2}}$$

The **Debye–Waller factor** is related to the rms amplitude A of the thermal vibration of the atoms in the unit cell; it is a function of the temperature:

$$B = 8\pi \overline{u^2}$$

$$\sqrt{\overline{u^2}} = A_{rms}$$

At the diffraction condition, the temperature factor thus becomes: $e^{-\frac{4\pi^2 \overline{u^2}}{d^2}}$

The temperature factor is ≈ 1 at small diffraction angles, getting larger as 2θ is increased (high angles). The effect is to decrease the value of the integrated intensity J. The peak heights become smaller, but not wider. The lost intensity goes into raising the background level.

At high temperatures (where quantum effects can be ignored):

$$KE_{atom} = \frac{3}{2}kT$$

$$= \frac{1}{2}m\overline{v^2}$$

The displacement u of an atom undergoing oscillatory motion is:

$$u = u_o \sin \omega t \qquad \text{displacement}$$

$$v = \omega u_o \cos \omega t \quad \text{velocity}$$

$$\overline{v^2} = \omega^2 \overline{u^2}$$

$$\frac{1}{2}m\omega^2 \overline{u^2} = \frac{1}{2}kT \qquad \begin{array}{l}\text{½ for equipartition of energy in}\\ \text{one direction}\end{array}$$

$$\overline{u^2} = \frac{kT}{m4\pi^2}\frac{1}{f^2}$$

$$\boxed{B = \frac{2kT}{m}\frac{1}{f^2}} \quad B \text{ is a linear function of } T.$$

1.4.22 X-Ray Analysis of a Diffraction Pattern – Cubic

For a cubic crystal structure with lattice parameter a, the spacing d between planes is given by:

$$d = \frac{a}{\sqrt{h^2 + k^2 + l^2}}$$

Planes of smaller Miller indices (100, 110, 111) give diffraction lines at low values of 2θ where: $\lambda = 2d\sin\theta$

Eliminating d, we have:

$$\sin\theta = \frac{\lambda}{2a}\sqrt{h^2 + k^2 + l^2}$$

and

$$\sin^2\theta = \frac{\lambda^2}{4a^2}\,h^2 + k^2 + l^2$$

h, k and l are positive integers, and so for any pair of diffraction lines:

$$\frac{\sin^2\theta_1}{\sin^2\theta_2} = \frac{h_1^2 + k_1^2 + l_1^2}{h_2^2 + k_2^2 + l_2^2} = \frac{N_1}{N_2}$$

Our diffraction patterns consist of a number of peaks at various values of 2θ. We begin by considering any pair of peaks, say the first two. The ratio of $\sin^2\theta_1$ and $\sin^2\theta_2$ is found and expressed in terms of the lowest possible two integers N_1 and N_2. (hkl) are then determined by inspection.

Information about the lattice type can be found in the h, k and l values obtained.

FCC: all odd or all even
BCC: sum of h, k and l is even
Primitive cubic: none of the above

Here, for the first two peaks, we have:

$$\frac{\sin^2\theta_1}{\sin^2\theta_2} = \frac{0.02501}{0.06612} = \frac{3}{8}$$

$$h_1^2 + k_1^2 + l_1^2 = 3$$

$$\therefore (hkl) = (111)$$

$$h_2^2 + k_2^2 + l_2^2 = 8$$

$$\therefore (hkl) = (220)$$

Measured from diffraction pattern. Note that low (hkl) found at low 2θ

Peak	2θ	$\sin^2\theta$	$\dfrac{\sin^2\theta_1}{\sin^2\theta_2}$	N	hkl
1	18.2	0.0250		3	111
			0.3782		
2	29.8	0.0661		8	220
			0.7212		
3	35.25	0.0916		11	311
			0.6886		
4	42.8	0.1334		16	400
			0.6664		
5	53.1	0.1999		24	422
			0.8889		
6	56.6	0.2247		27	333
			0.8424		
7	62.2	0.2668		32	400
			0.8029		
8	70.4	0.3322		40	620
			0.9281		
9	73.5	0.3579		43	533

Analysis of all the lines (e.g., N_3/N_4, N_4/N_5, N_5/N_6... etc.) reveals that all the (hkl) sets are either all odd or all even, indicating an FCC structure. The lattice parameter can be calculated using any one set of (hkl):

$$a = \frac{\lambda}{2\sin\theta}\sqrt{h^2 + k^2 + l^2} \approx 8.44 \text{ Å}$$

1.5 Thermal Properties of Solids

Summary

$$g(\omega) = \frac{L}{2\pi}\frac{1}{v}$$

Density of states (1D)

$$g(\omega) = \frac{3V}{2\pi^2}\frac{\omega^2}{v^3}$$

Density of states (3D)

$$C = 3R$$

Dulong–Petit law

$$C \approx Nk\left(\frac{\hbar\omega}{kT}\right)^2 e^{-\hbar\omega/kT}$$

Einstein model

Debye model

$$C = \frac{3V}{2\pi^2 v^3}\frac{\hbar^2}{kT^2}\int\limits_0^{\omega_D} \frac{\omega^4 e^{\hbar\omega/kT}}{\left(e^{\hbar\omega/kT}-1\right)^2}\,d\omega$$

$$\alpha = \frac{\partial x}{\partial T} = \frac{3gk}{4a^2}$$

Thermal expansion coefficient

$$K = \frac{1}{3}Cv\lambda$$

Thermal conductivity (non-metal)

$$K = \frac{1}{3}\left(\frac{\pi^2 Nk^2 T}{2E_F}\right)v_F \lambda_F$$

Thermal conductivity (metal)

1.5.1 Density of Vibrational States

A travelling wave, such as the propagation of a small elastic deformation in a solid, can be represented by a trigonometric cosine function:

$$u(x,t) = A\cos(\omega t - kx + \phi) \quad \text{where } k = \frac{2\pi}{\lambda}$$

For mathematical convenience, a wave can also be expressed in complex exponential form: $u(x,t) = Ae^{i(\omega t - kx + \phi)} = Ae^{i(-kx + \phi)}$

We can set $t = 0$ when we are not interested in time-dependent effects. This expression represents one possible solution to the general wave equation. Boundary conditions determine the precise nature of the solution. For example, imposition of the periodic boundary conditions means that the **wave number** k can only take on certain values:

$$k = 2\frac{n\pi}{L} \quad n = 0, \pm1, \pm2, \pm3, \pm4...$$

where L is the characteristic length over which the boundary conditions apply. In one dimensional k-space, we can represent the allowed values of k as:

wave travelling to the left $\frac{2\pi}{L}$ ⟶| |⟵   wave travelling to the right

Each point in k-space represents a particular mode of vibration in the solid.

Now, $v = f\lambda = \omega\dfrac{\lambda}{2\pi} = \dfrac{\omega}{k}$

thus, $\boxed{\omega = vk}$

Dispersion relation for elastic wave in a solid. Slope is equal to the velocity of sound in the solid.

This is called a **dispersion relation** and connects the frequency of the wave to the wave number.

If L is relatively large, then there are many possible modes or states that are very finely spaced. For an interval dk in k-space, the number of modes n within this interval is:

$$n = \frac{L}{2\pi}dk = \frac{L}{2\pi}\frac{1}{v}d\omega \quad \text{for the case of a linear dispersion relation.}$$

The number of modes per unit of frequency is called the **density of states** $g(\omega)$. For one dimension:

$$\boxed{g(\omega) = \frac{L}{2\pi}\frac{1}{v}} \quad \text{constant, independent of } \omega$$

In three dimensions, of volume V, where we may have one longitudinal and two transverse modes of vibration with the same wave number, the density of states is not a constant with ω but is expressed:

$$\boxed{g(\omega) = \frac{3V}{2\pi^2}\frac{\omega^2}{v^3}}$$

1.5.2 Classical Harmonic Oscillator Model

In **classical thermodynamics**, we might treat each vibrating atom or molecule as a **harmonic oscillator** (i.e., the atoms or molecules are connected to other atoms or molecules by linear springs and are undergoing **simple harmonic motion**). All the atoms in the solid have the same frequency of vibration. The energy of vibration of an individual atom is:

$$E = \frac{1}{2}mv^2 + \frac{1}{2}kx^2$$
$$= \frac{1}{2}mv^2 + \frac{1}{2}m\omega^2 x^2$$

where m is the mass, x is the displacement from the equilibrium position and v is the instantaneous velocity.

The total **internal energy** U in a crystalline solid is N times the average energy E of the atoms where N is the total number of atoms in the lattice. At any one point in time, there will be a continuum of velocities of atoms in the solid which is described by the **Maxwell velocity distribution**.

For atoms in a solid, there is only vibrational motion in three possible directions (no rotation or translations – or else it wouldn't be a solid). Averaging over all possible velocities, Boltzmann found that the total energy U is given by:

$$U = 3kT$$

Note that the total energy U does not depend upon the frequency of oscillation.

For a mole of atoms, the total energy is:

$$U = 3N_A kT$$
$$= 3RT \quad \text{Since } R = N_A k$$

R = 8.3145 J mol⁻¹ K⁻¹

But, $\dfrac{dU}{dT} = C$ the molar specific heat.

Thus: $\boxed{C = 3R}$ **Dulong–Petit law**

That is, this classical model predicts that the **molar specific heat** for all solids is a constant equal to $3R$. This is true for most solids at reasonably high temperatures (but not near the melting point) but is not observed to hold at low temperatures close to absolute zero, where C actually decreases.

1.5.3 Einstein Harmonic Oscillator Model

The inability of the **Dulong–Petit law** to agree with experimental results for measurements of specific heat at low temperatures led Einstein to propose that the internal energy of oscillation is quantised in nature according to:

$$E = h\nu = \hbar\omega$$

where ω is the resonant frequency of the oscillating atom. In his model, Einstein proposed that all the atoms were identical, and had the same resonant frequency, and were vibrating independently of each other in the sense that the motion of one atom did not affect the motion of any neighbouring atom. Because of the quantisation of energy, each of the atomic oscillators thus had to have a minimum of $E = h\nu$ energy to contribute to the specific heat of the solid. Further, any energy greater than this would have to occur in integral steps of $h\nu$. Energies thus start at 0 (no contribution to the specific heat) and continue upwards in steps $nh\nu$ where n is 0, 1, 2, ...

The significance of this concept is that N, the number of atomic oscillators that are actually oscillating (and contributing to C) falls off at low temperatures and thus causes C to fall below that predicted by the classical treatment.

By including the quantisation of energy to **Boltzmann statistics**, the average energy per atom is expressed:

$$E = \frac{\hbar\omega}{e^{\hbar\omega/kT} - 1}$$

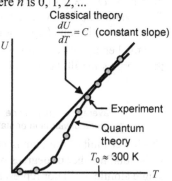

Classical theory
$\dfrac{dU}{dT} = C$ (constant slope)

Experiment

Quantum theory

$T_0 \approx 300$ K

At high temperatures, the total energy U approaches the classical result: $U = kT$. At low temperatures: $U \approx \hbar\omega e^{-\hbar\omega/kT}$

If the oscillator frequency ω is known, then the predicted values of C agree with those observed in experiments where at low temperatures:

$$\boxed{\frac{dU}{dT} = C \approx Nk\left(\frac{\hbar\omega}{kT}\right)^2 e^{-\hbar\omega/kT}}$$

Calculations show that the oscillator frequency at 300K is on the order of 10^{12} Hz.

It is important to note that in the **Einstein model**, it is assumed that all the atomic oscillators are vibrating at the same frequency. Increasing the temperature increases the amplitude of vibration of any one atom, but any change in energy must occur in discrete steps. The discreteness of the steps depends upon what the resonant frequency of vibration is. The atoms do not have the same energy as one another (velocity and amplitude of vibration). The distribution of energies is described by Boltzmann statistics.

1.5.4 Debye Lattice Model

The **Einstein model** does not correctly predict the specific heat at very low temperatures because of its simplifying assumption that the atoms in the solid are vibrating independently of each other at the same resonant frequency. In a crystalline solid, large regions of the crystal vibrate in unison at a much lower frequency than the characteristic vibrational modes of the individual atoms. Such low frequency, long wavelength "acoustic" vibrations move with the speed of sound in the solid. Thus, the total energy of vibration within the solid has a contribution from these low frequency, long wavelength vibrations. At very low temperatures, the energies associated with these vibrations may be such that: $\hbar\omega < kT$

Debye assumed that these low frequency vibrations possessed a linear dispersion $\omega = vk$ (as for an elastic continuum) and that, in contrast to the Einstein model, there were a range of frequencies present which satisfied certain periodic boundary conditions set by the dimensions of the crystal.

The total energy of vibration U for the crystal lattice is found by summing the energies for all the possible modes of vibration according to:

$$U = \int E(\omega)g(\omega)d\omega \quad \text{where} \quad E = \frac{\hbar\omega}{e^{\hbar\omega/kT} - 1}$$

average energy of one mode of vibration or state   density of states $g(\omega) = \dfrac{3V}{2\pi^2}\dfrac{\omega^2}{v^3}$

The lower limit of this sum is $\omega = 0$. Debye determined that the upper limit is the total number of degrees of freedom for the entire solid. That is, each single atomic oscillator represents one possible mode of vibration with three degrees of freedom. For one mole of atoms, the upper limit to the number of modes is $3N_A$. The **Debye frequency** ω_D is thus found from:

$$\int_0^{\omega_D} g(\omega)d\omega = 3N_A \quad \text{where it is found that:} \quad \boxed{\omega_D = v\left(6\pi^2\frac{N_A}{V}\right)^{1/3}}$$

In k-space, this is a sphere of radius k_D. The specific heat C is found by differentiating U with respect to T to give:

$$\boxed{C = \frac{3V}{2\pi^2 v^3}\frac{\hbar^2}{kT^2}\int_0^{\omega_D}\frac{\omega^4 e^{\hbar\omega/kT}}{\left(e^{\hbar\omega/kT} - 1\right)^2}d\omega}$$

At very low temperatures, this expression yields $C \propto T^3$ in accordance with experimental observations. It is at these very low temperatures that these long wavelength **acoustic waves**, treated as if there were in an elastic continuum, contribute to the specific heat.

1.5.5 Phonons

Debye's simplification of acoustic waves as if they were in a homogeneous elastic continuum is accompanied by the requirement of the periodic boundary conditions which permit only certain values of the **wave number** k.

Einstein quantised lattice vibrations in terms of energy:

$$E = n\hbar\omega$$

The energy unit of quantisation is called a phonon. For a wave of average energy E, then n, the number of phonons present, is:

$$n = \frac{1}{e^{\hbar\omega/kT} - 1}$$

The high frequency atomic vibrations produce **optical phonons**, whereas the low frequency oscillations, where quantisation arises due to the periodic boundary conditions, are termed **acoustic phonons**.

Phonons are waves emanating from harmonic oscillators in a crystal lattice. If everything were perfect, phonons would not interact with each other but simply combine and separate via the principle of superposition when they crossed. Imperfections in both the harmonic nature of the oscillation, and geometry of the crystal cause colliding phonons to scatter. Such scattering results in the phenomenon of **thermal conductivity**, while the anharmonic oscillations give rise to **thermal expansion**.

The de Broglie relation permits a momentum to be associated with phonons.

$$p = \frac{h}{\lambda} = \hbar k$$

The momentum associated with a phonon has particular importance for the electronic properties of **indirect band gap semiconductors**.

Energy changes in a crystal (such as a consequence of changing the temperature of the crystal) can only occur in multiples of $h\nu$, which corresponds to one phonon at a time. Phonons are created by raising the temperature, and are destroyed by lowering the temperature. The required energy balance is achieved by conduction of heat into and out of the solid. Application of the **Schrödinger equation** for a harmonic oscillator for modes of vibration of a crystal yields the requirement that the energy of vibration is:

$$E = \left(n + \frac{1}{2}\right)\hbar\omega$$

which has the interesting consequence of predicting a **zero-point energy** of vibration of $E_o = 1/2\,\hbar\omega$ when $n = 0$ (at absolute zero). This has particular relevance to the phenomenon of **superconductivity**.

1.5.6 Lattice Waves

The models considered so far have ignored the presence of atoms in the crystal lattice and have instead been concerned with the motion of waves in an elastic continuum. The significance of this is shown in the **dispersion relation**. In the case of long wavelength vibrations, the dispersion is linear because the wavelength of the phonons is very much greater than the periodicity of the lattice.

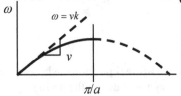

When the wavelength becomes significant with respect to the lattice parameters, the presence of atoms at the lattice sites scatters the wave and decreases their velocity. This causes the dispersion to deviate from the linear form.

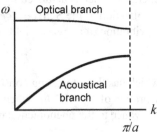

The motion of a single line of atoms is analysed as if they were connected by linear springs of force constant α (called the **interatomic force constant**) and where each atom has the same mass. When this is done, it is found that the dispersion relation is a sinusoid with period $2\pi/a$ in k–space.

$$\omega = \frac{2}{a}v\sin\frac{ka}{2}$$

In the long wavelength limit:
$$\sin\frac{ka}{2} \approx \frac{ka}{2} \text{ hence: } \omega = \frac{2}{a}v\frac{ka}{2} = vk$$

As the wavelength becomes shorter, we eventually reach a condition where at $k = \pi/a$, there is no wave velocity – a **standing wave** in the lattice. The portion of the curve between $-\pi/a$ and π/a is called the **Brillouin zone**. When a line of two atoms per unit cell (where there are two masses contributing to the same wave motion) is analysed in the same way, it is found that there are two solutions, an upper **optical branch** and a lower **acoustical branch**. These two solutions represent vibrations whereby in the optical branch, the two atoms move in opposite directions to each other (in and out from the centre of mass of the unit cell), while in the acoustical branch, atoms move in the same direction with respect to the centre of mass of the unit cell.

The dispersion curve for the optical branch lies above that of the acoustic branch. When the two masses are equal, the dispersion curves touch at the edge of the Brillouin zone. When the two masses are different, there is a band of frequencies that represent disallowed modes of vibration of the lattice.

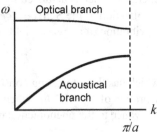

1.5.7 Thermal Expansion

Consider the variation in potential between two atoms or molecules in a solid. Thermal energy exists in the form of oscillations about the equilibrium position. As the temperature is raised, the amplitude of the oscillations is increased and if the oscillations were symmetrical, around the equilibrium position, the average position of the atoms relative to one another would remain unchanged. The dimensions of the solid would thus not change with temperature.

The potential within which the atoms move is in fact not symmetrical, and so the oscillations of the atoms are not harmonic. Thus, as the temperature is raised, the average position of the atoms becomes more "off-centre" and the solid expands. It is interesting to note that observed values of thermal expansion, in relative terms, are only about 10% of the amplitude of thermal vibration.

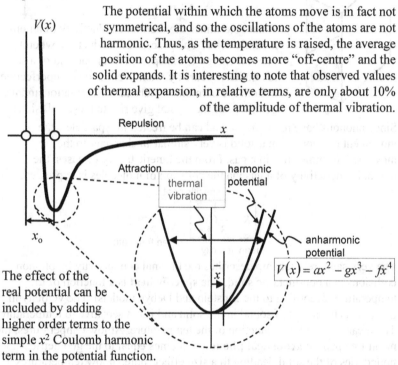

$$V(x) = ax^2 - gx^3 - fx^4$$

The effect of the real potential can be included by adding higher order terms to the simple x^2 Coulomb harmonic term in the potential function.

It is desired to obtain the average displacement $\bar{x}$ from the equilibrium position x_o as a function of temperature T. This is done using **Boltzmann statistics**, whereupon it is found that:

$$\bar{x} = \frac{3}{4}\frac{g}{a^2}kT$$

Thermal expansion coefficient: $\alpha = \dfrac{\partial x}{\partial T} = \dfrac{3gk}{4a^2}$

1.5.8 Thermal Conductivity (Non-Metals)

For a unit cross sectional area, the rate of heat flow ($J\ s^{-1}$) through a solid of
length L is usually expressed using the heat conduction formula:

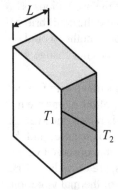

$$\dot{Q} = K\frac{(T_1 - T_2)}{L}$$

thermal conductivity
$W\ m^{-1}\ K^{-1}$

In a metal, it is the
movement of free
electrons that
dominates the heat
conduction process.

What determines the thermal conductivity of a non-
metallic solid? The process of conduction, whereby
thermal energy is transported from one end of a solid
to another, occurs by phonon collisions. Imperfections
in both the harmonic nature of the oscillation and the
geometry of the crystal give rise to these collisions.

Since phonons have momentum, and can be treated like particles, the
movement of phonons in a solid is very similar in character to the
movement of molecules in a gas. From the kinetic theory of gases, the
thermal conductivity of a gas via the collision of molecules is expressed:

specific heat

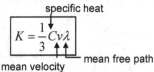

$$K = \frac{1}{3}Cv\lambda$$

mean velocity ———— mean free path

The very same relationship describes the thermal conductivity by phonon
collisions in a non-metallic solid. The **specific heat** is a function of the
temperature (according to the Einstein and Debye models). The velocity in
this case is the speed of sound in the solid and is not sensitive to temperature.
The mean free path is a function of the temperature. At low temperatures, the
mean free path of **acoustical phonons** becomes equal to the geometrical
boundaries of the solid, leading to a size effect (note: at low temperatures
there are very few optical phonons). At high temperatures, atomic
displacements have a large amplitude and the anharmonic character
oscillations of **optical phonons** becomes more pronounced, thus increasing
the amount of scattering that occurs, thereby decreasing λ. The overall result
is that at very low temperatures, the dependence of the thermal conductivity
on temperature is governed by the specific heat (T^3). At high temperatures,
the thermal conductivity is inversely proportional to T due to the influence of
anharmonic oscillations and subsequent scattering of optical phonons.

1.5.9 Thermal Conductivity (Metals)

In metals, the contribution of phonons to thermal conductivity is very much less than that of the free electrons in the conduction band. In a metal, the free electrons at the hot end of a solid have more energy than those at the cold end. Free electrons travel from the hot to the cold end at the same rate as those travelling from the cold to the hot end; but because those that travel from the hot to the cold end have more energy, a net energy transfer takes place between the hot and cold ends of the solid.

The significance of this is that it is only those energetic electrons at the **Fermi surface** which contribute to the net transfer of energy and thus determine the thermal conductivity of the solid. These electrons have an average velocity:

$$\overline{v^2} = 2\frac{E_F}{m}$$

The motion of free electrons is similar to that of molecules in a gas, and, as in the case of an insulator, the thermal conductivity K is expressed:

$$K = \frac{1}{3}Cv\lambda$$

where C is the specific heat, v is the velocity and λ is the mean free path of the particles, in this case, the conduction electrons. When the appropriate quantities are used, the thermal conductivity of the conduction electrons is:

$$\boxed{K = \frac{1}{3}\left(\frac{\pi^2 Nk^2 T}{2E_F}\right)v_F\lambda_F}$$

This term represents the specific heat of conduction electrons (that is, only those electrons within the range kT of the Fermi surface) where N is the number of free electrons per unit volume in the solid.

Note that the specific heat of the conduction electrons is linearly dependent upon the temperature T.

The thermal conductivity K is usually expressed in terms of the electrical conductivity σ through the **Lorentz number** L:

$$\boxed{K = LT\sigma} \quad \text{where } L = \frac{\pi^2}{3}\left(\frac{k}{q_e}\right)^2 \text{ is a constant}$$

1.6 Mechanical Properties of Solids

Summary

$F = kx$ Hooke's law

$\dfrac{\sigma}{\varepsilon} = E$ Young's modulus

$U = \dfrac{1}{2} kx^2$ Strain energy

$\Delta U = 2\gamma$ Surface energy

$\dfrac{\pi \sigma_a^2 c}{E} \geq 2\gamma$ Griffith energy balance

1.6.1 Hooke's Law

Consider the shape of the force law between two atoms or molecules in more detail. Its shape resembles that of a sine wave in the vicinity of the force maximum.

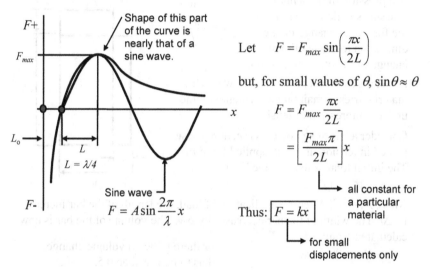

Let $\quad F = F_{max} \sin\left(\dfrac{\pi x}{2L}\right)$

but, for small values of θ, $\sin\theta \approx \theta$

$$F = F_{max}\frac{\pi x}{2L}$$

$$= \left[\frac{F_{max}\pi}{2L}\right]x$$

all constant for a particular material

Thus: $\boxed{F = kx}$

for small displacements only

F may be expressed in terms of force per unit area (or **stress**), which is given the symbol σ.

$$\sigma = \frac{\sigma_{max}\pi}{2L}x$$

Let the fractional change in displacement from the equilibrium position (the **strain**) be given by:

$$\varepsilon = \frac{x}{L_o}$$

Substituting for x and transferring ε gives:

$$\frac{\sigma}{\varepsilon} = \left[\frac{L_o\pi\sigma_{max}}{2L}\right] \quad \textbf{Hooke's law}$$

All material properties

$$= E$$

Young's modulus or "stiffness"

1.6.2 Poisson's Ratio

It is observed that for many materials, when stretched or compressed along the length within the elastic limit, there is a contraction or expansion of the sides as well as an extension or compression of the length.
Poisson's ratio is the ratio of the fractional change in one dimension to the fractional change of the other dimension.

$$v = \frac{\dfrac{\Delta w}{w}}{\dfrac{\Delta l}{l}}$$

Poisson's ratio is a measure of how much a material tries to maintain a constant volume under compression or tension.

Consider a bar of square cross section $w \times w$ placed in tension under an applied force F. The initial total volume of the bar is:

$$V_1 = A_1 l$$

where $A_1 = w^2$. After the application of load, the length of the bar increases by Δl. The width of the bar decreases by Δw. The volume of the bar is now calculated from:

$$V_2 = (l + \Delta l)(w - \Delta w)^2$$

$$= l(1 + \varepsilon)w^2 \left(1 - \frac{\Delta w}{w}\right)^2$$

$$= l(1 + \varepsilon)A_1(1 - v\varepsilon)^2$$

$$\approx A_1 l(1 + \varepsilon - 2v\varepsilon) \quad \substack{\text{since} \\ \varepsilon^2 \ll 1}$$

The change in volume is thus:

$$V_2 - V_1 \approx A_1 l - A_1 l(1 + \varepsilon - 2v\varepsilon)$$

$$= A_1 l\varepsilon(1 - 2v)$$

For there to be no volume change, v has to be less than 0.5. $v > 0.5$ implies that the volume decreases with tension, an unlikely event. When $v = 0.5$, there is no volume change and the contraction in width is quite pronounced (e.g., rubber). When $v = 0$, the volume change is the largest and there is no perceptible contraction in width. Most materials have a value of v within the range 0.2 to 0.4.

When the material contracts inward (a so-called **plane stress** condition) under an applied tensile stress σ_T, there is no sideways stress induced in the material. If the sides of the material are held in position by external forces or restraints (**plane strain**), then there is a stress σ induced, the value of which is given by $\sigma = v\sigma_T$.

In terms of stresses and strains, in plane strain conditions (sides held in position), there is an effective increase in the stiffness of the specimen due to the induced sideways stresses. **Hooke's law** becomes $\sigma = \dfrac{E}{1 - v^2}\varepsilon.$

1.6.3 Surface Energy

Forces between atoms or molecules take the form of a repulsion that is very strong at short distances and an attraction which diminishes in strength with larger distances. Atoms and molecules take up an equilibrium position where the **repulsive** and **attractive forces** are balanced.

Consider two molecules in a solid, one on the surface, and another in the interior. **Long-range** attractive forces F_A have a resultant R_A zero on A, downwards on B. **Short-range** repulsive forces F_R have a resultant R_R zero on A, upwards on B.

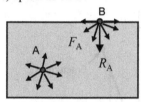

But, attractive forces are *long-range* forces repulsive forces are *short range*; thus, a molecule B at the surface feels an attraction from the molecule at layers 1 and 2 since attractive forces are long range, $R_A = F_{A1} + F_{A2}$. However, the molecule at the surface only feels the repulsion from the molecule directly beneath it ($R_R = F_{R1}$) since repulsive forces are short range. Thus, to counterbalance all the "extra" attractive forces from the deeper molecules, the surface molecule has to move downward and closer to layer 1 since the repulsive force increases with decreasing distance.

The effect is to create a **surface tension** in the surface. This is most often pronounced in liquids, but exists in solids although the deflections involved are very much smaller in magnitude.

The added downward attraction that affects atoms on the surface of a solid represents energy. Creation of a surface, such as cleaving a solid, requires not only the energy to break the bonds, but also the energy to create the two new surfaces that result. The energy required to create the two new surfaces is thus:

$$\Delta U = 2\gamma$$

where γ is the **surface energy** of the solid (≈ 1 J m^{-2} for most solids). If two like surfaces are in contact within a medium of different surface energy, then the energy associated with the interface is:

$$\gamma = \gamma_1 + \gamma_2 - 2\sqrt{\gamma_1 \gamma_2}$$

1.6.4 Brittle Fracture

Hooke's law applies to the linear elastic region. When load is removed, the body returns to its original shape. If the body is stretched beyond the elastic limit, it will only partially return to its original shape and thus acquire a **permanent set**.

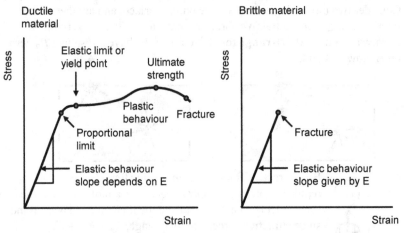

When a solid is stretched or compressed by the application of an applied force, the application of the force F through a distance dx requires work to be done on the system.

$$dU = Fdx$$
$$F = kx$$
$$U = \int kx\,dx$$
$$= \frac{1}{2}kx^2 \qquad \text{Strain potential energy}$$

When a solid fractures, the stored **strain potential energy** is converted into heat, kinetic energy, plastic deformation and surfaces – that is, the surface energy of the cracked parts. **Brittle materials** generally shatter into many surfaces which soak up the stored strain energy released during fracture. **Ductile materials** tend to absorb the stored strain energy in the fracture surfaces and also **plastic deformation** inside the material.

1.6.5 Griffith Energy Balance

Consider the application of a uniform tensile stress to a solid. The distribution of stress within the material is uniform as shown. A crack of length c is then introduced into the material. Stresses are relieved in the material in the vicinity of the crack, and lines of tension tend to concentrate around the crack tip. The sharper the tip of the crack, the greater is the stress concentration at the tip. If the applied stress is increased, then the crack may grow in size if the rate of release of strain energy with respect to crack length is sufficient to create new crack surfaces. This is expressed mathematically by the **Griffith criterion** for crack growth:

$$\frac{dU_s}{dc} \geq \frac{dU_\gamma}{dc}$$

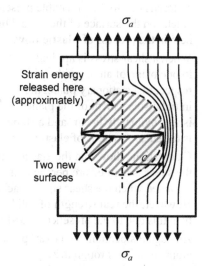

The strain energy released by introducing a double-ended crack of length $2c$ in an infinite plate of unit width under a uniformly applied stress σ_a is

$$U_s = \frac{\pi \sigma_a^2 c^2}{E}$$

The total surface energy for *two* surfaces of unit width and length $2c$ is

$$U_\gamma = 4\gamma c$$

The Griffith criterion is thus: $\boxed{\dfrac{\pi \sigma_a^2 c}{E} \geq 2\gamma}$

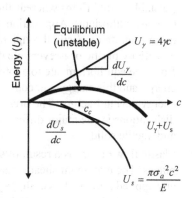

For a given applied stress, a crack will not extend unless the Griffith criterion is met. That is, a crack needs to be greater than a critical length c_c in order for there to be sufficient strain energy released to create two new surfaces.

1.6.6 Dislocations in Solids

Solids are in general, either brittle or ductile. **Brittle solids** transfer strain energy into surfaces and are generally weak in comparison to ductile materials due to the inevitable presence of stress-concentrating flaws or cracks on the surface of the solid. **Ductile solids** absorb strain energy as heat in the process of **plastic flow**.

When a shear stress is applied to one plane of atoms with respect to a stationary plane, the initial displacement of the planes 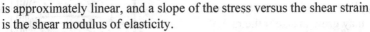 is approximately linear, and a slope of the stress versus the shear strain is the shear modulus of elasticity.

Calculations show that the elastic limit (when the atoms in the plane to which the stress is applied are perched on top of the stationary atoms underneath) is achieved when the shear stress reaches about 1/6 of the shear modulus. However, the real strength of solids is usually in orders of magnitude less than this due to the presence of **dislocations** in the crystal lattice.

An edge dislocation is an extra plane of atoms running through the crystal structure. This can happen in shear when some atoms slip to a new equilibrium position but others do not. The endpoints of the extra plane are called a **dislocation line** or **dislocation axis**.

The atoms at the sensitive end of the dislocation exist at a potential hill (unstable equilibrium) whereas the other atoms are in potential wells (stable equilibrium). Application of a stress causes the dislocation to move or translate fairly easily through the crystal structure. This results in what is essentially plastic or irreversible deformation of the solid as atoms shift from one equilibrium site to another. Dislocation movement requires energy and this energy can come from released strain energy during fracture. The presence and motion of dislocations is one major difference between brittle and ductile solids.

Plastic flow occurs as a result of shear across crystalline planes. The maximum shear stress is at an angle of 45° to any applied normal stress. Slippage usually therefore manifests itself as slip lines at 45° as shown:

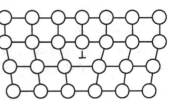

Part 2

Dielectric Properties of Materials

2.1 Dielectric Properties

Summary

$$\mu = Qd \qquad \text{Dipole moment}$$

$$\mu_E = \alpha E \qquad \text{Dipole moment}$$

$$\alpha = \alpha_e + \alpha_i + \alpha_d \qquad \text{Polarisability}$$

$$\varepsilon_r = \frac{C}{C_o} = \frac{V_o}{V} \qquad \text{Relative permittivity}$$

$$\varepsilon = \varepsilon_r \varepsilon_o \qquad \text{Permittivity}$$

$$P = N\mu_E \qquad \text{Polarisation}$$

$$\varepsilon_r = 1 + \frac{N\mu_E}{\varepsilon_o E} \qquad \text{Relative permittivity}$$

$$= 1 + \frac{N\alpha}{\varepsilon_o}$$

$$\frac{\varepsilon_r - 1}{\varepsilon_r + 2} = \frac{N\alpha}{3\varepsilon_o} \qquad \text{Clausius–Mosotti equation}$$

$$P = (\varepsilon_r - 1)\varepsilon_o E \qquad \text{Susceptibility}$$

$$= \chi \varepsilon_o E$$

$$\varepsilon_r = 1 + \frac{P}{\varepsilon_o E} \qquad \text{Relative permittivity}$$

$$\varepsilon^* = \varepsilon' - j\varepsilon'' \qquad \text{Complex relative permittivity}$$

2.1.1 Electric Charge

Electrical (and magnetic) effects are a consequence of a property of matter called **electric charge**. Experiments show that there are two types of charge that we label **positive** and **negative**. Experiments also show that unlike charges attract and like charges repel.

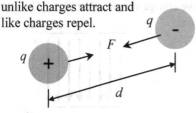

Experiments also show that the magnitude of the force between two charges is proportional to the magnitude of the charges and inversely proportional to the square of the distance between them. This is Coulomb's law.

$$F \propto \frac{q_1 q_2}{d^2}$$

$$= k \frac{q_1 q_2}{d^2}$$

k is a constant of proportionality, the nature of which we will investigate very soon.

Now,

1. imagine that one of the charges q_2 is hidden from view;
2. the other charge still experiences the Coulomb force and thus we say it is acted upon by an **electric field**;
3. if a **test charge** experiences a force when placed in a certain place, then an electric field exists at that place. The direction of the field is taken to be that in which a positive test charge would move in the field.

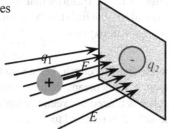

$$F = k \frac{q_1 q_2}{d^2}$$

$$\text{Let } E = k \frac{q_2}{d^2}$$

$$\text{thus } F = q_1 E$$

Note: the origin of the field E may be due to the presence of many charges but the magnitude and direction of the resultant field E can be obtained by measuring the force F on a single test charge q.

2.1.2 Electric Flux

An electric field may be represented by lines of force. Consider the presence of free or excess charges on a pair of parallel plates. For each free positive charge on the plate, we can imagine a line of force extending from this charge to the corresponding free negative charge on the other plate. The total number of lines is called the **electric flux** and is proportional to the amount of excess charge on one of the plates. The number of lines per unit cross-sectional area is the **electric flux density**, which can be expressed in terms of Coulombs per square meter and is given the symbol D.

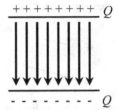

We may now ask, "What is the force on a test charge placed between the plates if we know the total free charge on the plates?" It is reasonable to conclude that the more free charges there are, the greater the force on the test charge. Thus, the force depends upon D. Is D then the same as E (the strength of the electric field that appears in Coulomb's law)? No. The actual value of the force experienced by a test charge for a given value of D also depends upon the medium, or material, through which the lines of force pass. Some materials are more effective at transmitting the lines of force than others. Thus, the value of E (the quantity that determines the resulting Coulomb force F) is proportional to D, which has been multiplied by an "effectiveness factor" which we will express as:

$$E = \left(\frac{1}{\varepsilon}\right)D \qquad \varepsilon_o = 8.85 \times 10^{-12}\,F\,m^{-1}$$

E is the **electric field strength**. For a vacuum, ε is written ε_o. For a material, ε is larger than ε_o. This means that for a given value D, the field E is reduced when a material is placed between the plates. When a *conducting* material is inserted, the field E is reduced to zero (ε is infinite and the "effectiveness factor" is zero). When an insulator is inserted, we call the insulator a **dielectric**; the field E is reduced by an amount which depends upon the material property ε for that insulator. We can see that the effectiveness of the material at producing a field strength E depends inversely upon ε, which is called the **permittivity**. D is the cause, E is the result. The force experienced by a test charge for a given value of D depends only on E, which in turn depends upon the nature of the material through ε.

Are the lines of force for an electric field lines for E or D? Both. The spacing of the lines indicates both the flux density and also the field strength. The effectiveness of the material in transforming D into a force-producing E depends upon ε for that material.

2.1.3 Dipole

1. Consider two opposite charges separated by a distance d.

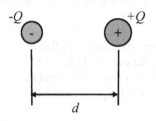

The two charges may be separated by their being fixed in position in a molecule, held apart by some external force, or any other reason. The important thing is that the two charges are kept separated.

2. Draw a circle around the two charges:

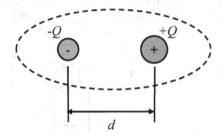

The entity enclosed by the circle is called a **dipole**. That is, a dipole consists of an assembly of two opposite charges separated by a distance d.

3. A net electric field exists between the two charges in a dipole. (Note, if the two charges were coincident, then the radial field from each would add up to zero net field. Since they are separated, then there is a net field between them.)

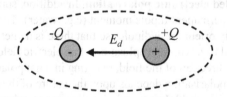

*An electric **dipole** produces an **electric field**.*

In the diagram here, only one field line is shown connecting the two charges. For single isolated charges, field lines would emanate radially in all directions from each charge and the resulting field is the vector sum of the individual fields.

Dipoles are responsible for the **dielectric behaviour** of materials (i.e., the response of materials in an electric field).

2.1.4 Polarisation

Consider an atom or molecule in a non-polar insulating material (i.e., where there are no existing permanent dipoles or free electrons).

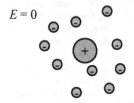

$E = 0$

$E = 0$

With no externally applied E field, the centre of charge of electrons corresponds with the centre of charge of protons.

When an electric field E is applied as shown below, the centre of charge of electrons moves left, the centre of charge of protons moves right and the two centres of charge are then separated by distance d. The atom or molecule becomes **polarised** in the presence of E.

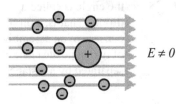

$E \neq 0$

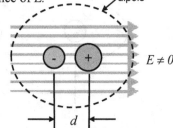

dipole

$E \neq 0$

d

*An **electric field** can thus produce an electric **dipole**.* The field created by the dipole interacts with the externally applied field so that the dipole tends to be aligned with the external field.

The example above is called **electronic polarisation**. In addition, some molecules already have a permanent dipole moment (e.g., water). Thermal agitation tends to randomly orient these dipoles so that there is no net polarisation of the material as a whole. Application of an electric field E tends to align the dipoles in the direction of the field, resulting in a net polarisation. In general, the amount of polarisation depends upon the nature of the material and the strength of the applied field. In general, the alignment tendency from the applied field is in competition with the randomising influence of temperature and so the amount of polarisation depends on temperature.

The field E may not necessarily be fixed in magnitude or direction. A sinusoidal **oscillating (AC) field** would result in the direction of the dipoles tending to follow the field in a periodic way, resulting in periodic movement of the atoms or molecules.

2.1.5 Dipole Moment

A dipole consists of a separation of opposite charges by a distance d. A dipole can exist on its own as a permanent entity (as in the case of a polar molecule) or be created by an external electric field E.

When the charge Q (Coulombs) is multiplied by the distance d (meters) between them, then this quantity is called the electric **dipole moment** μ:

The dipole moment μ is a **vector** pointing in the direction d.

$\mu \longrightarrow$

$\mu = Qd$

—— magnitude of (one of) the charges

distance between the centre of charge from $-Q$ to $+Q$

The **dipole moment** is a measure of how polarised an atom or molecule is. Different types of atoms or molecules polarise more easily than others in the presence of a field E. The amount of polarisation depends on the **polarisability** α of the material and the strength of the applied field E. Therefore, it is reasonable to write:

$$\mu_E = \alpha E$$

Note: μ is a vector. The subscript denotes the component of the dipole moment in the direction of E.

The units of dipole moment are usually given as "**debyes**" for convenience. The dipole moment for water is about 1.9 debyes. 1 debye = 1 × 10²⁹ Cm

It is important to note that the dipole moment is not the same thing as a mechanical moment or torque. A mechanical moment exists when the dipole is placed in a field and attempts to align itself with the field.

The potential **energy of a dipole**, oriented at an angle θ to an external field, is given by:

$$U = QV_2 - QV_1$$
$$= Q(V_2 - V_1)$$
$$E = -\frac{(V_2 - V_1)}{d\cos\theta}$$
$$(V_2 - V_1) = -Ed\cos\theta$$
$$U = -QdE\cos\theta$$
$$= -\mu_E E\cos\theta$$

$\mu_E = \mu\cos\theta$

2.1.6 Examples of Polarisability

Polarisation in a dielectric may occur due to several mechanisms all of which may occur to some extent depending on the material:

1. **Electronic polarisability** α_e

   Small induced dipole moment arising from difference in the net centres of nucleus and electrons in an atom.

   $\varepsilon_r \approx 2-4$

   The entities with circles around them are **dipoles**.

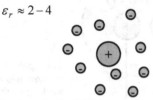
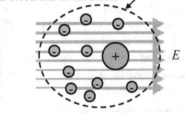

2. **Ionic polarisability** α_i

   Dipole moment created by shift of positive ions with respect to negative ions in unit cell.

   $\varepsilon_r \approx 6-10$

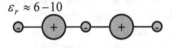
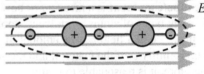

   When field is applied, movement of ions produces a net dipole moment in a unit cell and hence a net polarisation.

3. **Dipolar polarisability** α_d

   Net dipole moment created by alignment of molecule with external field due to presence of internal permanent dipoles from geometrical structure of molecule.

   $\varepsilon_r \approx 20-100$

   Water molecule has a permanent dipole moment.

   CO_2 molecule has no permanent dipole moment. No dipolar polarisation can occur.

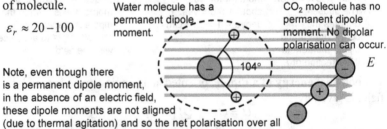

   Note, even though there is a permanent dipole moment, in the absence of an electric field, these dipole moments are not aligned (due to thermal agitation) and so the net polarisation over all molecules in the material is zero. When a field E is applied, molecular dipoles tend to align themselves with the field and there exists a net polarisation.

Total **polarisability** α

$$\alpha = \alpha_e + \alpha_i + \alpha_d$$

All materials have α_e, ionic materials have α_e, α_i while dipolar materials have α_e, α_i and α_d. The relative contribution of each to the total polarisability depends upon the frequency of oscillation of the field E, e.g., at high frequencies, dipolar molecules cannot oscillate fast enough to keep up with the change in direction of E so the α_d contribution falls.

2.1.7 Conductor in an Electric Field

To understand the behavior of a dielectric in an electric field, it is useful to examine what happens within a **conductor** in an electric field. Consider a parallel plate capacitor which has been charged with a voltage V_o.

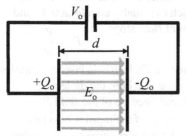

An electric field E_o is established between the plates by the build-up of charge on each plate. Current in the circuit drops to zero when the potential due to the charges Q_o becomes equal to V_o.

Next, imagine a conductor is inserted between the plates so that it fills the space between the plates but does not actually touch the plates.

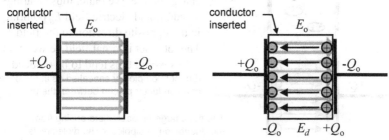

Since there are many **free electrons** in the conductor, they all tend to accumulate towards the end of the conductor nearest to the +ve plate of the capacitor. This movement of electrons results in a build-up of -ve charge on one side of the conductor accompanied by a build-up of +ve charge on the other end. Movement of charge happens until the electric potential between the two ends of the conductor becomes equal to the potential V_o between the plates. The build-up of internal charge within the conductor results in an electric field E_d within the conductor pointing in the opposite direction to the original field between the plates. *The total net field E within the conductor is thus brought to zero.*

This can happen with a conductor because there are many mobile charge carriers (electrons) that are free to move. Movement can occur until the material within the conductor becomes completely polarised in the presence of the external field. In an insulator, as we shall see, only restricted movement of charge within the material can occur and only partial polarisation occurs.

2.1.8 Dielectric in an Electric Field

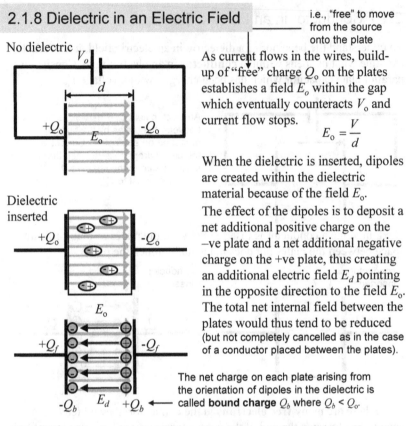

i.e., "free" to move from the source onto the plate

As current flows in the wires, build-up of "free" charge Q_o on the plates establishes a field E_o within the gap which eventually counteracts V_o and current flow stops.

$$E_o = \frac{V}{d}$$

When the dielectric is inserted, dipoles are created within the dielectric material because of the field E_o.

The effect of the dipoles is to deposit a net additional positive charge on the −ve plate and a net additional negative charge on the +ve plate, thus creating an additional electric field E_d pointing in the opposite direction to the field E_o. The total net internal field between the plates would thus tend to be reduced (but not completely cancelled as in the case of a conductor placed between the plates).

The net charge on each plate arising from the orientation of dipoles in the dielectric is called **bound charge** Q_b where $Q_b < Q_o$.

BUT! In the situation shown above, where there is a fixed voltage V_o across the plates ($V_o = E_o d$ and d is also fixed), E_o between the plates must be maintained and this is done by drawing extra "free" charge into the plates from the source V_o to counteract the field E_d from bound charge. The total **free charge** on each plate is now: $Q_f = Q_o + Q_b$. Since $C = Q_f/V$, the **capacitance** of the plates has increased due to this extra charge. That is, the presence of a dielectric increases the capacitance of a parallel plate capacitor.

If the plates were charged with V_o and the voltage source removed, and then the dielectric inserted, then the total net charge on each plate would be reduced to $Q_o - Q_b$, the total net field between the plates would be reduced to $E = E_o - E_d$ and because d is a constant, the voltage across the plates would be reduced from V_o to V. The total *free* charge on the plates is now $Q_f = Q_o$ and so, since $C = Q_f/V$, the capacitance is thus increased by the presence of the dielectric.

2.1.9 Permittivity

Consider a parallel plate capacitor charged to a voltage V_o and then the voltage source is removed. A voltmeter placed across the capacitor terminals will register the voltage V_o used to charge the capacitor.

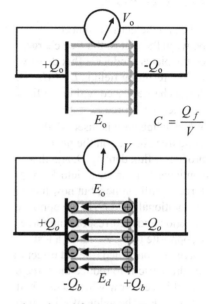

$$C = \frac{Q_f}{V}$$

When a dielectric is inserted, dipoles are created within the dielectric material because of the field E_o leading to a net additional charge Q_b (i.e., in addition to Q_o) being deposited on each plate.

Since the total net charge on each plate is now reduced in magnitude, the voltage recorded on a voltmeter across the plates is reduced to V and the total average field between the plates is reduced from E_o to E. But, the total _free_ charge on each plate (i.e., the charge initially drawn from the voltage source) remains the same: $Q_f = Q_o$.

Since the distance d is a constant, then the ratio of voltage V_o/V gives the relative change in capacitance of the plates. By definition, the **dielectric constant**, or **relative permittivity**, ε_r, of an insulating material is the capacitance with the dielectric material inserted divided by the capacitance when the plates are in a vacuum.

$$\varepsilon_r = \frac{C}{C_o} = \frac{V_o}{V}$$

Expressed in terms of the fields E_o and E, we have:

$$\varepsilon_r = \frac{E_o}{E}$$

The **relative permittivity** is a macroscopic measure of the amount by which a dielectric material is polarised by an electric field. The permittivity of an actual dielectric is found from:

$$\varepsilon_r = \frac{\varepsilon}{\varepsilon_o}$$

Permittivity of free space (vacuum)
8.85×10^{-12} F m^{-1}

2.1.10 D and E

When no dielectric is placed between the plates, the free charges on the plates produce a flux density D. The resulting field is $E_o = 1/\varepsilon_o\, D$. A test charge placed within the field experiences a force $F_o = qE_o$.

When a dielectric is inserted, the same free charges on the plate produce a flux density D, and this continues to produce a field E_o, and force F_o still applies, but now there is an additional field E_d in the opposite direction and a corresponding force F_d opposite to F_o. As far as a test charge is concerned, the net effect of the dielectric is to reduce the force on it as if it were now within a net field $E = E_o - E_d$. The value of E is found from:

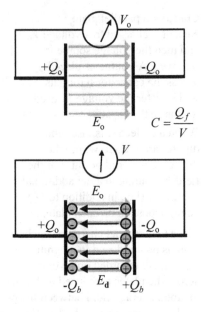

$$C = \frac{Q_f}{V}$$

$$E = \frac{1}{\varepsilon}D$$

$$D = \varepsilon E = \varepsilon_o E_o$$

The difference in the net field between the plates before and after insertion of the dielectric is:

$$E_d = E_o - E$$

and arises due to polarisation of the dielectric. Expressed in terms of D, we have: $E = E_o - E_d$

$$= \frac{\varepsilon}{\varepsilon_o}E - E_d$$

$$\frac{\varepsilon}{\varepsilon_o}E = E + E_d$$

flux density arising from the polarisation field E_d

$$D = \varepsilon E = \varepsilon_o E + (\varepsilon_o E_d)$$

actual flux density from the free charges on the plates

flux density that would have to exist with no dielectric if the net field was really E instead of E_o

The flux density D is sometimes called the **electric displacement**.

The electric field E is sometimes called the **electric field intensity, electric field strength**.

2.1.11 Energy of the Electric Field

An electric field can do work and so has potential energy. It is of interest to determine what energy is available at any point within the field. As an example, we consider the uniform field between the parallel plates of a capacitor.

In a circuit with a **capacitor**, energy is expended by the voltage source as it forces charge onto the plates of the capacitor. When fully charged, and disconnected from the voltage source, the voltage across the capacitor remains. The stored electric potential energy within the charged capacitor may be released when desired by discharging the capacitor. To find the energy stored in a capacitor, we can start by considering the power used during charging it.

Uniform electric field

$$E = \frac{Q}{\varepsilon_0 A}$$

Now, $V = Ed$

thus $V = \frac{Q}{\varepsilon_0 A} d$

Power $P = vi$

$i = \dfrac{dq}{dt}$

Lower case letters refer to instantaneous quantities.

$Pdt = vdq = U$

$$U = \int_0^Q vdq = \int_0^Q \frac{q}{C}dq \quad \text{Energy}$$

$$= \frac{1}{2}\frac{Q^2}{C}$$

$$\boxed{U = \frac{1}{2}CV^2}$$

Energy stored in the electric field of a parallel plate capacitor charged to voltage V

$$C = \frac{Q}{V}$$

$$= Q\frac{\varepsilon_0 A}{Qd}$$

$$C = \varepsilon_0 \frac{A}{d}$$

2.1.12 Polarisation

For a fixed voltage V across the plates, by Gauss' law, we have:

$$8.85 \times 10^{-12} \text{ F m}^{-1}$$

$$EA = \frac{Q_f}{\varepsilon} \longrightarrow \varepsilon = \varepsilon_r \varepsilon_o$$

$$= \frac{Q_o + Q_b}{\varepsilon} \overset{\text{extra free charge brought}}{\underset{\text{the bound charge (fixed } V)}{\text{onto the plate to counteract}}}$$

$$\varepsilon E = \frac{Q_o}{A} + \frac{Q_b}{A} = D \quad \begin{array}{l}\text{electric field}\\\text{intensity}\end{array}$$

Let $P = \dfrac{Q_b}{A}$

$$= \frac{Q_b d}{Ad} \quad \begin{array}{l}\text{Multiply top \&}\\\text{bottom by } d\end{array}$$

$$= \frac{Q_b d}{V}$$

$$P = \frac{n\mu_E}{V} \quad \text{Since } n\mu_E = Q_b d$$

$$\boxed{P = N\mu_E} \quad \begin{array}{l}n = \text{number of}\\\text{dipoles present}\end{array}$$

↳ number of dipoles per unit volume (dipole density)

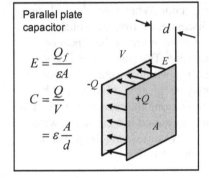

Parallel plate capacitor

$$E = \frac{Q_f}{\varepsilon A}$$

$$C = \frac{Q}{V}$$

$$= \varepsilon \frac{A}{d}$$

The quantity P is the **electric dipole moment per unit volume** of the dielectric material and is called the **polarisation**.

Thus

$$\varepsilon E = \frac{Q_o}{A} + P$$

$$= \varepsilon_o E + P$$

↓ electric field in the dielectric after insertion of dielectric

The polarisation is a macroscopic quantity. If we know the polarisation, we don't have to worry about adding up all the individual dipole moments within the material. Note that the bound charge arises at the surfaces of the dielectric where they meet the plates. Inside, the charges cancel.

The **polarisability** α of a material describes *how easily* an atom or molecule can become polarised in an electric field. The **polarisation** P described *how much* polarisation has taken place per unit volume of material when field E has been applied.

2.1.13 Dielectric Susceptibility

In the case of a parallel plate
capacitor, we have:

$$\varepsilon E = \varepsilon_0 E + P$$

Therefore $\varepsilon_r \varepsilon_0 E = \varepsilon_0 E + P$ since $\varepsilon = \varepsilon_r \varepsilon_0$

$$\varepsilon_r = 1 + \frac{P}{\varepsilon_0 E}$$

$$= 1 + \chi$$

χ is called the **dielectric susceptibility**
and is a measure of the amount to
which a dielectric is polarised by the
field E over that for a vacuum ε_0.

It is easy to show that:

and also

$$\boxed{\begin{aligned} \varepsilon_r &= 1 + \frac{N\mu_E}{\varepsilon_0 E} \\ &= 1 + \frac{N\alpha}{\varepsilon_0} \end{aligned}}$$

$$P = (\varepsilon_r - 1)\varepsilon_0 E$$
$$= \chi \varepsilon_0 E$$
$$\varepsilon_r = 1 + \frac{P}{\varepsilon_0 E}$$

where the macroscopic quantity ε_r, the **relative permittivity**, is expressed
in terms of the **polarisability** α of the material and the number of dipoles
per unit volume N (the last two of which are microscopic quantities).

The **relative permittivity** is a macroscopic measure of the dielectric and
optical properties of a material, where the material may contain various
atoms and molecules of possibly various individual polarisabilities.

The electric flux density D can be now expressed in terms of the
polarisation of the dielectric over which it acts:

$$D = \varepsilon E = \varepsilon_0 E + P$$

2.1.14 Clausius–Mosotti Equation

In solids, experiments show that the relation

$$\varepsilon_r = 1 + \frac{N\mu_E}{\varepsilon_0 E} \longrightarrow \text{external field}$$

does not hold exactly because the local field experienced by a single dipole is not equal to the applied external field E but is a little larger than this due to the proximity of other nearby dipoles.

Lorentz calculated that the local field E_{loc} was equal to:

$$E_{loc} = E + \frac{P}{3\varepsilon_0}$$

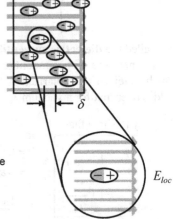

E_{loc}

Substituting back, we obtain a new expression for P which includes the local field correction:

$$P = N\mu_E \quad \text{\small N is the number of dipoles per unit volume}$$

$$= N\alpha E_{loc}$$

$$P = N\alpha\left(E + \frac{P}{3\varepsilon_0}\right)$$

$$P = \frac{N\alpha E}{1 - \dfrac{N\alpha}{3\varepsilon_0}} \longleftarrow \text{\small The denominator here is the local field correction}$$

Now, $\varepsilon_r = 1 + \dfrac{P}{\varepsilon_0 E}$

$$= 1 + \frac{N\alpha E}{\left(1 - \dfrac{N\alpha}{3\varepsilon_0}\right)\varepsilon_0 E}$$

⇩

$$\boxed{\frac{\varepsilon_r - 1}{\varepsilon_r + 2} = \frac{N\alpha}{3\varepsilon_0}}$$ **Clausius–Mosotti equation**

Note: At first it might be thought that the presence of other dipoles tends to cancel the applied field so that the local field would be less than E. There is indeed an overall depolarisation field created and this reduces the *average* field strength between the plates, but *very near* to the molecules, the *local* field is actually increased due to the presence of nearby dipoles. This is usually significant for solids in which atoms or molecules are relatively closely spaced.

This expression for ε_r takes into account the local field correction, which is important for solids (large N), but less important for gases (small N).

2.1.15 Complex Permittivity

An ideal dielectric is non-conducting, but there are always losses in a real dielectric (e.g., **leakage**). For a capacitor, this is represented by a **loss resistor** R_L.

Without the loss resistor, the capacitance of the capacitor (with a dielectric) is:

$$C = \varepsilon_r \left[\varepsilon_0 \frac{A}{d} \right]$$

where $C_0 = \varepsilon_0 \frac{A}{d}$

With the loss resistor, the impedance of the capacitor is found from:

$$V = IZ$$

where $\frac{1}{Z} = \frac{1}{R_L} + j\omega C$

Now, the resistance R_L can be expressed in terms of the **conductivity** α of the dielectric:

resistivity
$$R_L = \rho \frac{d}{A}$$

$$\frac{1}{R_L} = \sigma \frac{A}{d}$$

conductivity

Note: R is small when A/d is large. That is, for a given d, large capacitors have greater losses.

The impedance of the capacitor becomes:

$$\frac{1}{Z} = \sigma \frac{A}{d} + j\omega C$$

$$= \sigma \frac{A}{d} + j\omega \varepsilon_r C_0$$

$$= j\omega C_0 \left(\varepsilon_r - j\sigma \frac{A}{d} \frac{1}{\omega C_0} \right)$$

$$= j\omega C_0 \left(\varepsilon_r - j\sigma \frac{A}{d} \frac{1}{\omega \varepsilon_0} \frac{d}{A} \right)$$

$$= j\omega C_0 \left(\varepsilon_r - j \frac{\sigma}{\omega \varepsilon_0} \right)$$

$$= j\omega C_0 (\varepsilon' - j\varepsilon'') \text{ where } \varepsilon' = \varepsilon_r$$

$$\varepsilon'' = \frac{\sigma}{\omega \varepsilon_0}$$

Complex relative permittivity

$$\boxed{\varepsilon^* = \varepsilon' - j\varepsilon''}$$

This term reflects the resistive or conduction losses in a capacitor. It may also contain terms associated with frictional losses in dipolar materials.

The **phase angle** δ is a measure of the quality of a dielectric. The **loss factor** D is given by:

$$D = \tan \delta = \frac{1}{R_L \omega C} = \frac{\varepsilon''}{\varepsilon'}$$

- Ideal capacitor: $D = 0$
- High quality dielectric: $D \approx 10$
- Lossy capacitor: $D \approx 0.05$

Time constant τ of a dielectric is:

$$\tau = R_L C = \rho \varepsilon_0 \varepsilon_r$$

When R_L and C are constant, the D is inversely proportional to ω.

τ is independent of the dimensions of the capacitor and only depends on the nature of the dielectric.

2.2 Polarisability

Summary

$$\alpha_e = 4\pi\varepsilon_o R^3$$ Electronic polarisability

$$\alpha_i = \frac{Q_{eff}^2}{m\omega_o^2}$$ Ionic polarisability

$$\langle \mu_E \rangle = \frac{\mu^2}{3kT}E$$ Dipole moment

$$\alpha_d = \frac{\mu^2}{3kT}$$ Dipolar polarisability

$$P(t) = P_\infty + (P_s - P_\infty)\left(1 - e^{-t/\tau}\right)$$ Dielectric loss

$$P(t) = (\varepsilon^* - 1)\varepsilon_o E(t)$$

$$\varepsilon' = \varepsilon_\infty + \frac{\varepsilon_s - \varepsilon_\infty}{1 + \omega^2\tau^2}$$ Debye equations

$$\varepsilon'' = \frac{\varepsilon_s - \varepsilon_\infty}{1 + \omega^2\tau^2}\omega\tau$$

$$\varepsilon'' = \frac{\sigma}{\omega\varepsilon_o} + \frac{(\varepsilon_s - \varepsilon_\infty)}{1 + \omega^2\tau^2}\omega\tau$$ Dipolar dispersion

$$\varepsilon_r(\omega) = \varepsilon_\infty + \frac{\varepsilon_s - \varepsilon_\infty}{1 - \omega^2/\omega_i^2}$$ Ionic dispersion

$$\varepsilon_\infty(\omega) = n^2(\omega) = 1 + \frac{NZe^2/\varepsilon_o m}{1 - \omega_o^2/\omega^2}$$ Electronic dispersion

2.2.1 Electronic Polarisability

Consider an atom as consisting of a
positively charged nucleus $+Ze$
surrounded by a negatively charged
sphere of radius R *which is completely
filled with a uniform distribution of
electron charge* $-Ze$. When a local
field E is present, the off-centre shift
results in an opposing depolarisation
field E_d over the distance δ.

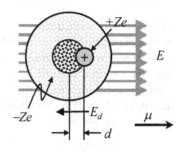

Equilibrium is established when the field E_d, due to the electrons enclosed
by the smaller sphere of radius δ, balances the external field E. By **Gauss'
law**, we have:

Note: In this book:
$$e = 1.6 \times 10^{-19} \text{C}$$
$$-q_e = -1.6 \times 10^{-19} \text{C}$$

$$\varepsilon_o \int_{S_d} EdA = q_{enclosed}$$

surface of
sphere d
$$= \frac{4/3\,\pi d^3}{4/3\,\pi R^3}(-Ze)$$

$$\varepsilon_o E 4\pi d^2 = \frac{d^3}{R^3}(-Ze)$$

but $\mu_E = -Zed$

thus $\mu_E = 4\pi\varepsilon_o R^3 E$

$$= \alpha_e E$$

$$\boxed{\alpha_e = 4\pi\varepsilon_o R^3}\quad \textbf{Electronic polarisability}$$

A typical value of R would be 1×10^{-10} m (1 Å or 0.1 nm). Thus:

$$\alpha_e = 4\pi\left(8.85 \times 10^{-12}\right)\left(1 \times 10^{-10}\right)^3$$
$$= 1.1 \times 10^{-40}\,\text{F m}^2$$

A value for ε_r can be found from the Clausius–Mosotti equation:

$$\frac{\varepsilon_r - 1}{\varepsilon_r + 2} = \frac{N\alpha}{3\varepsilon_0}$$

$$= \frac{5 \times 10^{28}\left(1.1 \times 10^{-40}\right)}{3\left(8.85 \times 10^{-12}\right)}$$

letting $N = 5 \times 10^{28}$ m^{-3}

$$\varepsilon_r \approx 1.8$$

2.2.2 Ionic Polarisability

In ionic polarisation, a dipole moment is created by the shift of positive ions with respect to negative ions in a unit cell. The dipole moment is given by the magnitude of the effective charge Q_{eff} times the distance d:

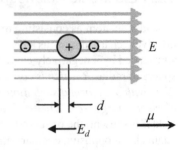

$$\mu_E = Q_{eff}\delta$$

The **ionic bonds**, for small displacements, act like springs of a stiffness k. Thus, the off-centre movement occurs until force equilibrium is achieved such that:

┌── spring constant of the ionic bond

$$Q_{eff}E = kd$$

but $\mu_E = \dfrac{Q_{eff}^2}{k}E$

Q_{eff} is normally less than Ze due to distortion of electron charge.

thus $\boxed{\alpha_i = \dfrac{Q_{eff}^2}{k}}$ **Ionic polarisability**

It is interesting to determine the **resonant frequency** of the system since this will determine its response to an oscillating applied field. Assuming simple harmonic motion, we have a resonant frequency given by:

$$\omega_o = \sqrt{\frac{k}{m}}$$

where m is the effective mass of the system given by: $\dfrac{1}{m} = \dfrac{1}{m_{-ion}} + \dfrac{1}{m_{+ion}}$

The **ionic polarisability** can thus be expressed:

$$\boxed{\alpha_i = \dfrac{Q_{eff}^2}{m\omega_o^2}}$$

Generally speaking, the ionic polarisability is not a function of the temperature of the material.

2.2.3 Dipolar Polarisability

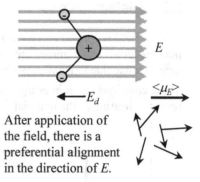

Dipolar polarisation occurs when there is a preferential orientation of polar molecules within the material as a result of the applied field E.

Before the application of the field, the dipole moments are oriented at random.

After application of the field, there is a preferential alignment in the direction of E.

To determine the polarisation, we require the average net dipole moment in the direction of the field E for all the molecules.

Now, the distribution of energies of molecules at some temperature T is given by the **Boltzmann distribution**:

$$f(W)dW = Ae^{-W/kT}dW$$

$k = 1.38 \times 10^{-23}$ J K^{-1}
T = absolute temperature

where $\int f(W)dW = 1$ where the integral is taken over all the possible energies

The energy of a single dipole oriented at angle θ to the applied local field E is:

$$W = -\mu E \cos\theta$$

$$\mu_E = \mu \cos\theta$$

and so: $dW = \mu E \sin\theta$

The Boltzmann distribution becomes:

$$f(W)dW = Ae^{-\mu E\cos\theta/kT}\mu E\sin\theta\,d\theta \quad \text{where } \theta \text{ goes from 0 to 180°}$$

$$= f(\theta)$$

The average value of μ_E is: $\langle\mu_E\rangle = \dfrac{\displaystyle\int_0^\pi \mu\cos\theta f(\theta)d\theta}{\displaystyle\int_0^\pi f(\theta)d\theta}$

In solids, the dipoles usually are not able to freely rotate and are instead confined to discrete orientations. This leads to a modification of the factor 1/3 for α_d.

the molecule's average net dipole moment in the direction of E

$$\langle\mu_E\rangle = \frac{\mu^2}{3kT}E$$

the molecule's permanent dipole moment

Dipolar polarisability $\boxed{\alpha_d = \dfrac{\mu^2}{3kT}}$

that is, the dipolar polarisability varies inversely with the absolute temperature T.

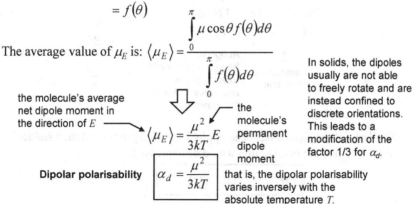

2.2.4 Dielectric Loss

When a field is applied to a dielectric, the response of the dipoles (for a polar material) is not immediate since the dipole takes a certain amount of time to rotate due to friction by collisions with other molecules and inertia.

These collisions result in energy drawn from the applied field, which results in heating of the material. This is called **dielectric loss**.

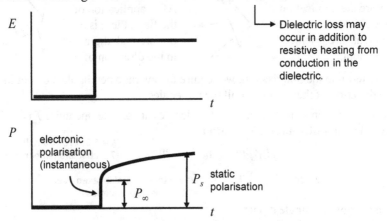

Dielectric loss may occur in addition to resistive heating from conduction in the dielectric.

If the field is applied at $t = 0$, then

$$P(t) = P_\infty + (P_s - P_\infty)(1 - e^{-t/\tau})$$

For polar liquids, $\tau \approx 10^{-9}$ to 10^{-11} sec.
For free ions, $\tau \approx 10^{-5}$ to 10^{-2} sec.

dielectric **relaxation time**

Now, in general: $P = (\varepsilon_r - 1)\varepsilon_0 E$

thus, $P_s = (\varepsilon_s - 1)\varepsilon_0 E$

and $P_\infty = (\varepsilon_\infty - 1)\varepsilon_0 E$

instantaneous polarisation

At optical frequencies $\varepsilon_r = n^2$ where n is the **refractive index** of the material.

$P = N\mu_E$

$\varepsilon_r = 1 + \dfrac{N\mu_E}{\varepsilon_0 E}$

$= 1 + \dfrac{N\alpha}{\varepsilon_0}$

Similarly, if the field E has been applied for a long time, and then suddenly removed, then the polarisation would be expressed:

$$P(t) = P_\infty + (P_s - P_\infty)e^{-t/\tau}$$

2.2.5 Complex Permittivity

When an oscillating field is applied, the instantaneous response P_∞ to the field is due to **electronic polarisation** only since the electrons, at least up to the microwave and optical range of oscillation frequencies, are light enough to respond instantly to the changing direction of the field.

The dipolar contribution to the total polarisation is not instantaneous because it involves rotation of, and collisions between, molecules as the direction of the dipole attempts to follow the change in direction of the electric field. This results in the generation of heat – **dielectric loss**. That is, energy is drawn from the field E and is converted to heat within the dielectric.

Because of the non-instantaneous response, the total polarisation is thus not in phase with the applied field. The total polarisability is a complex quantity.

For a time-varying field $E(t)$, the total polarisation P therefore also varies with time according to:

$$P(t) = (\varepsilon * - 1)\varepsilon_o E(t)$$

When there is dielectric loss, relative permittivity is complex and the imaginary term contains information about these losses.

$$\varepsilon^* = \varepsilon' - j\varepsilon''$$

Note: Here we are talking mainly about losses due to dipole friction. Other losses may also occur (e.g., due to DC leakage) and these will also add as separate terms to the imaginary part of the complex permittivity.

We wish to determine an expression for the **complex permittivity** ε^* in terms of the instantaneous contribution ε_∞ and the total static value ε_s. Since the electronic polarisation is in phase with the field, we would expect ε_∞ to be contained within the real part of the complex permittivity ε'.

The difference between the static value and the instantaneous value, $\varepsilon_s - \varepsilon_\infty$. contains both real and imaginary components, the relative proportion being dependent upon the value of the **relaxation time** τ. For example, at $\tau = 0$, there would be of course no imaginary component and $\varepsilon' = \varepsilon_s$. The greater the relaxation time, the greater the magnitude of the imaginary component.

For an **alternating field**: $E(t) = E_o e^{j\omega t}$
The complex permittivity for the case of a dipolar material is given by:

$$\varepsilon^* = \varepsilon' - j\varepsilon'' = \varepsilon_\infty + \frac{\varepsilon_s - \varepsilon_\infty}{1 + j\omega\tau}$$

Separating out the real and imaginary components, we obtain:

$$\boxed{\varepsilon' = \varepsilon_\infty + \frac{\varepsilon_s - \varepsilon_\infty}{1 + \omega^2\tau^2}} \quad \text{and} \quad \boxed{\varepsilon'' = \frac{\varepsilon_s - \varepsilon_\infty}{1 + \omega^2\tau^2}\omega\tau} \quad \textbf{Debye's equations}$$

2.2.6 Dipolar Dispersion

The **complex permittivity** is expressed: $\varepsilon^* = \varepsilon' - j\varepsilon''$

where $\varepsilon' = \varepsilon_\infty + \dfrac{\varepsilon_s - \varepsilon_\infty}{1 + (\omega\tau)^2}$ and $\varepsilon'' = \dfrac{\varepsilon_s - \varepsilon_\infty}{1 + (\omega\tau)^2}\omega\tau$

The magnitude of the real and imaginary components is a function of frequency ω. This is called **dispersion**. Dispersion arises due to the non-zero **relaxation time** associated with the response of the dipoles to a changing electric field.

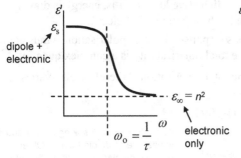

In general, the relaxation time τ is greater in solids compared to that in liquids. For solids, liquids and gases, the value of τ depends upon the shape of the molecule and the orientation with respect to the applied field. For a large collection of molecules (e.g., liquid water), there may be a **distribution of relaxation times** present. This influences the width of the peak in ε''.

Collision frequency (≈ 10.5 GHz in a water molecule. Note, this is *not* the frequency used in a microwave oven!)

The **phase angle** ϕ is a measure of the relative contributions of ε' and ε'' to the total complex permittivity: $\tan\phi = \dfrac{\varepsilon''}{\varepsilon'}$

When there is some **conductivity** present, then losses occur due to resistive or Joule heating such that the total imaginary component of relative permittivity is:

$$\varepsilon'' = \underbrace{\frac{\sigma}{\omega\varepsilon_0}}_{\substack{\text{resistive}\\\text{losses}}} + \underbrace{\frac{(\varepsilon_s - \varepsilon_\infty)}{1 + \omega^2\tau^2}\omega\tau}_{\substack{\text{frictional}\\\text{losses}}}$$

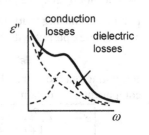

2.2.7 Ionic & Electronic Dispersion

Dispersion also arises in **ionic polarisation** in an oscillating field and is expressed:

$$\varepsilon_r(\omega) = \varepsilon_\infty + \frac{\varepsilon_s - \varepsilon_\infty}{1 - \omega^2/\omega_t^2}$$

↓

frequency of the optical phonon

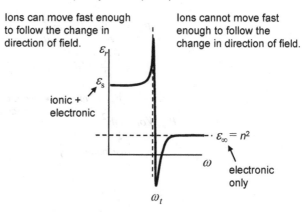

Ions can move fast enough to follow the change in direction of field.

Ions cannot move fast enough to follow the change in direction of field.

Dispersion in **electronic polarisation** occurs at very high frequencies.

$$\varepsilon_\infty(\omega) = n^2(\omega) = 1 + \frac{NZe^2/\varepsilon_0 m}{1 - \omega_0^2/\omega^2}$$

↓

Resonant frequency of electrons (of mass m subject to an electrostatic restoring force)

This is a very simplified expression applicable to an isolated atom and a single electron. In a solid, transitions between energy bands lead to very complex forms for dispersion and multiple resonant frequencies.

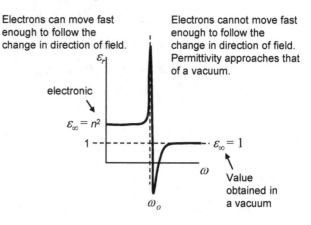

Electrons can move fast enough to follow the change in direction of field.

Electrons cannot move fast enough to follow the change in direction of field. Permittivity approaches that of a vacuum.

2.2.8 Power Dissipation

In the case of a parallel plate capacitor, the resistance loss term can be written:

$$\frac{1}{R} = \frac{\sigma A}{d}$$

The power dissipation is thus:

$$P_{av} = \frac{V_{rms}^2}{R} = V_{rms}^2 \sigma \frac{A}{d}$$

$$= \frac{V_{rms}^2}{d^2} \sigma A d$$

$$= V_{rms}^2 \varepsilon'' \omega C_o \longrightarrow \text{in the case of a parallel plate capacitor}$$

$$P_{av}/Vol = \sigma E_{rms}^2$$

$$\downarrow$$

AC conductivity

$$\varepsilon'' = \frac{\sigma}{\omega \varepsilon_o}$$

$$\sigma = \varepsilon'' \omega \varepsilon_o$$

For frictional losses,

$$P/Vol = \sigma E_{rms}^2$$

$$= \varepsilon'' \omega \varepsilon_o E_{rms}^2$$

$$= \frac{(\varepsilon_s - \varepsilon_\infty)\omega \tau}{1 + \omega^2 \tau^2} \omega \varepsilon_o E_{rms}^2$$

$$P_{max} \rightarrow \varepsilon_o \omega^2 (\varepsilon_s - \varepsilon_\infty)\frac{1}{\tau}$$

as $\omega \rightarrow \infty$

at $\omega = 1/3\tau$ $P = 0.1 P_{max}$

$\omega = 3\tau$ $P = 0.9 P_{max}$

2.3 Ferroelectric and Piezoelectric Materials

Summary

$\varepsilon_r = A \dfrac{P_s e^{-W/kT}}{DkT}$	Ferroelectric relative permittivity
$\varepsilon_r = \dfrac{A}{T - T_c}$	Curie–Weiss law
$\varepsilon^* = \varepsilon' - j\varepsilon''$	AC conductivity and relative permittivity
$\varepsilon'' = \dfrac{\sigma_{AC}}{\omega \varepsilon_o}$	
$\tan \delta = \dfrac{\sigma_{AC}}{\omega \varepsilon_o \varepsilon'}$	
$k^2 = \dfrac{U_e}{U_m}$	Piezoelectric coupling factor
$C_{HF} = \left(1 - k^2\right)C_{LF}$	Piezoelectric frequency response
$f_{\min} = \dfrac{1}{2\pi}\sqrt{\dfrac{1}{L_o C_o}}$	Piezoelectric low frequency resonance
$f_{\max} = \dfrac{1}{2\pi}\sqrt{\dfrac{C_o + C_1}{L_o C_o C_1}}$	Piezoelectric high frequency resonance
$k = \sqrt{1 - \left(\dfrac{f_{\min}}{f_{\max}}\right)^2}$	Piezoelectric coupling factor

2.3.1 Ferroelectricity

In a material such as **barium titanate** $BaTiO_3$, the Ti^{4+} ions may be in one of 6 minimum energy positions within the crystal lattice, all of which are slightly off-centre with respect to the ion charge distribution. This results in a static permanent dipole moment associated with the crystallographic unit cell, much like there is a permanent dipole in dipolar materials.

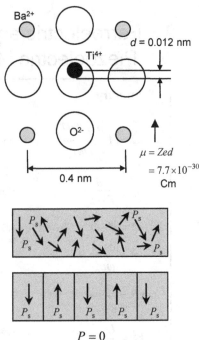

Ba^{2+}

$d = 0.012$ nm

Ti^{4+}

O^{2-}

0.4 nm

$\mu = Zed$
$= 7.7 \times 10^{-30}$ Cm

Initially, each individual dipole moment is oriented at random, but if the temperature is below the **Curie temperature** T_c (of about 380 K for $BaTiO_3$), the dipoles tend to interact with each other and line up in preferred orientations. It is as if the induced field from one dipole tends to preferentially induce the dipole in a neighbouring unit cell to align in the same direction.

$P = 0$

The overall result is a cooperative alignment of the dipoles associated with several crystallographic unit cells into a **domain**. Domains are regions within a crystal in which all the dipole moments lie in the same direction (i.e., all the Ti^{4+} ions in a domain have moved in the same direction). Over a particular domain, a **spontaneous net polarisation** P_s appears without the aid of an applied external field. Because of the presence of these domains, the material is said to be **ferroelectric**. In the absence of an externally applied field, the polarisation of the whole crystal P is initially zero since the domains themselves have a random orientation with respect to each other.

As T increases, cooperative alignment of dipoles diminishes until the dipoles become randomly oriented at T_c. The polarisation P_s associated with a domain drops to zero (although the individual unit cells still retain their off-centre displacement of ions).

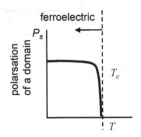

ferroelectric

P_s

polarisation of a domain

T_c

T

2.3.2 Relative Permittivity

The **relative permittivity** ε_r is a measure of how easily polarisation occurs in the presence of a field E. A high value of ε_r means that the induced polarisation from an applied field E is greater than that compared to a low value of ε_r for the same value of E.

In a ferroelectric material, such as $BaTiO_3$, the ease with which domains can align depends upon the ease with which Ti^{4+} ions can overcome the energy barrier and jump to a new position within the crystal lattice. When one considers the probability of a jump to and from available energy minima, the relative permittivity can be shown to be expressed:

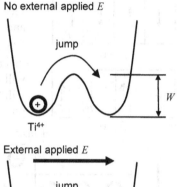

No external applied E

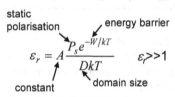

$$\varepsilon_r = A \frac{P_s e^{-W/kT}}{DkT} \quad \varepsilon_r \gg 1$$

static polarisation — energy barrier — constant — domain size

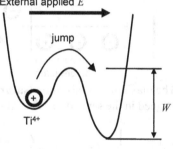

External applied E

In the presence of an external field, the jump is more favourable in the direction of E. Domains in the direction of E tend to grow in size.

As T increases, the ions are more easily able to overcome the energy barrier and so, in the presence of an external field E, domains can grow more easily as the temperature is raised. This leads to an increase in polarisation P of the crystal with increasing T. That is, ε_r increases with increasing temperature.

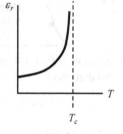

At the Curie temperature, there is a divergence of ε_r resulting from the influence of the local field on the mode of vibration of phonons within the crystal. A phase change in the crystal structure occurs at this temperature. Above T_c, the material is said to become **paraelectric** and ε_r then decreases with increasing temperature.

2.3.3 Ferroelectric Materials in an Electric Field

Below the **Curie temperature** T_c, spontaneous polarisations P_s occur in preferred orientations within the crystalline structure of the material, leading to **domains** of dipoles with the same orientation.

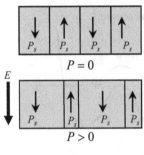

When no external field is applied, the **ferroelectric domains** are oriented at random and the total net polarisation P is zero.

When an external field E is applied, domains that are already aligned in the direction of E will grow and others not in the direction of E will become smaller. This leads to a net polarisation P within the material as a whole.

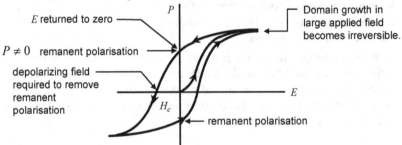

The maximum polarisation (**saturation**) occurs when all the domains are aligned in the same direction. At this point, $P = P_s$.

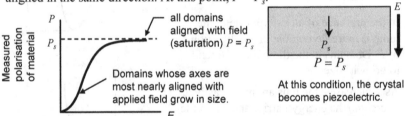

all domains aligned with field (saturation) $P = P_s$

Domains whose axes are most nearly aligned with applied field grow in size.

At this condition, the crystal becomes piezoelectric.

Measured polarisation of material

If the applied field is then reversed in direction and reduced to zero, it is found that there is some permanent net **remanent polarisation**.

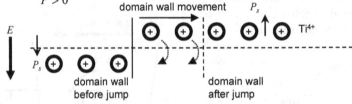

Domain growth in large applied field becomes irreversible.

E returned to zero

$P \neq 0$ remanent polarisation

depolarizing field required to remove remanent polarisation

remanent polarisation

2.3.4 Paraelectricity

Above the **Curie temperature** T_c, ionic dipoles in the material are at random orientation and the material is said to be in the **paraelectric phase**. The dielectric behaviour *is similar* to a dipolar material (i.e., because of the existence of the permanent ionic dipoles).

At the transition between the ferroelectric and paraelectric regimes, the relative permittivity ε_r is very high and so by the **Clausius–Mosotti equation**, we obtain:

$$\frac{\varepsilon_r - 1}{\varepsilon_r - 2} = \frac{N\alpha}{3\varepsilon_0} \approx 1$$

In a dipolar material, the polarisability is dependent upon the temperature such that:

$$\alpha = \frac{\mu^2}{3kT} \quad \longleftarrow \text{ the permanent dipole moment}$$

Assuming that the behaviour is like a dipolar material (i.e., because there are now permanent dipoles present), the **Curie temperature** can thus be found from:

$$\frac{N}{3\varepsilon_0} \frac{\mu^2}{3kT_c} \approx 1$$

$$T_c = \frac{N\mu^2}{9k\varepsilon_0}$$

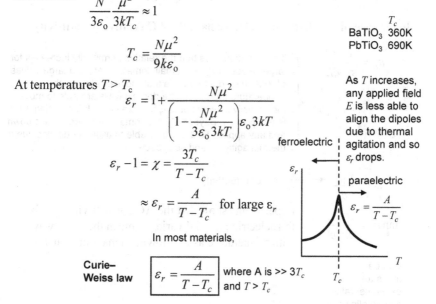

At temperatures $T > T_c$

$$\varepsilon_r = 1 + \frac{N\mu^2}{\left(1 - \frac{N\mu^2}{3\varepsilon_0 3kT}\right)\varepsilon_0 3kT}$$

$$\varepsilon_r - 1 = \chi = \frac{3T_c}{T - T_c}$$

$$\approx \varepsilon_r = \frac{A}{T - T_c} \text{ for large } \varepsilon_r$$

In most materials,

Curie– Weiss law

$$\boxed{\varepsilon_r = \frac{A}{T - T_c}}$$ where A is $\gg 3T_c$ and $T > T_c$

	T_c
BaTiO$_3$	360K
PbTiO$_3$	690K

As T increases, any applied field E is less able to align the dipoles due to thermal agitation and so ε_r drops.

ferroelectric

paraelectric

$$\varepsilon_r = \frac{A}{T - T_c}$$

T_c is a measure of the strength of the alignment of domains. The ratio $A/3T_c$ is a measure of how much alignment is left above T_c. Curie–Weiss law diverges as T approaches T_c.

2.3.5 AC Conductivity

When an oscillating field is applied to a **ferroelectric** material in the ferroelectric region, the domain walls migrate as the field changes direction. If we consider the transition from the point where $E = 0$ to E, then the domain wall shifts by an amount x.

The polarisation which then appears can be calculated from the total fractional change in the volume of domains now pointing in the direction of E:

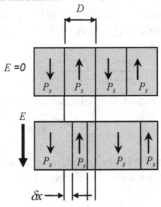

$$\delta P = P_s \left(2\frac{\delta x}{D} \right)$$

For large values of ε_r,

$$\varepsilon_r = \frac{1}{\varepsilon_0} \frac{P}{E}$$

In the case of an alternating field, we have the **AC relative permittivity** being:

$$\varepsilon_r = \frac{1}{\varepsilon_0} \frac{dP}{dE}$$

$$= \frac{2P_s}{D\varepsilon_0} \frac{dx}{dE}$$

In general,

$$\varepsilon^* = \varepsilon' - j\varepsilon''$$

$$\varepsilon'' = \frac{\sigma_{AC}}{\omega\varepsilon_0} \longleftarrow \text{ AC conductivity}$$

$$\tan\delta = \frac{\sigma_{AC}}{\omega\varepsilon_0\varepsilon'}$$

Can be measured experimentally and so allows σ_{AC} to be determined if we set $\varepsilon' = \varepsilon_r$.

This equation says that the relative permittivity increases for large values of P_s, and small domain sizes, and large values of dx/dE (i.e., domain walls that move easily). As the temperature increases, the domain walls are able to move more easily so we expect ε_r to increase with increasing T (at least up to T_c). Beyond T_c, the domain structure breaks down and any applied field E is less able to align the dipoles due to thermal agitation and so ε_r decreases.

Experiments show that the AC conductivity σ_{AC} for a ferroelectric material varies in much the same way with temperature as ε_r, showing a maximum at T_c.

2.3.6 Barium Titanate

Barium titanate is an important ferroelectric material. The dielectric behaviour can be studied by measuring its capacitance as a function of temperature.

In practice, ε_r can be calculated from the ratio of C/C_o where C is the measured capacitance of a parallel plate capacitor with the material to be tested as the dielectric between the plates and C_o is the capacitance calculated from the dimensions of the capacitor: $\varepsilon_o A/d$. For BaTiO$_3$, values of ε_r up to 10's of thousands are obtained near T_c.

Results for BaTiO$_3$ parallel plate capacitor for a diameter of 16 mm and a spacing of 2.2 mm at 1000 Hz as a function of temperature in C

Rearranging the Curie–Weiss law, we obtain:

$$\varepsilon_r T = \varepsilon_r T_c + A$$

slope ↑   ↑
intercept

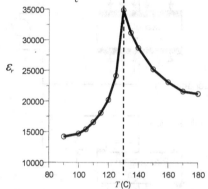

For the data shown above the Curie temperature shown here, we obtain T_c = 331 K from a least squares fitting to the data, a little lower than the expected value of 393 K.

Now,

$$T_c = \frac{N\mu^2}{9k\varepsilon_o}$$

If the lattice constant for the unit cell is given as 0.401 nm, then

$$N = \frac{1}{\left(0.402\times10^{-9}\right)^3} = 1.54\times10^{28}\,\text{m}^{-3}$$ No. dipoles per cell unit volume

Thus, the dipole moment μ for this experimental data is:

$$\mu = \frac{331(9)\left(8.85\times10^{-12}\right)\left(1.83\times10^{-23}\right)}{1.54\times10^{28}}$$ and $P_s = N\mu$

$$= 0.085 \text{ C m}^{-2}$$

$$= 5.51\times10^{-30} \text{ C m (Expected value 7.68 x 10}^{-30} \text{ C m)}$$

The offset of the Ti^{4+} ions can thus be calculated from:

$$\mu = Zed$$

$$5.51\times10^{-30} = 4\left(1.6\times10^{-19}\right)d$$

$$d = 0.0086 \text{ nm   (Expected value 0.012 nm)}$$

2.3.7 Piezoelectricity

Piezoelectricity is the electric field produced within certain materials when placed under a **strain**. Alternately, when an electric field is applied to a piezoelectric material, a strain is induced.

The piezoelectric effect occurs in **ferroelectric materials** when they are completely polarised (one single domain) and also in some non-ferroelectric materials such as quartz.

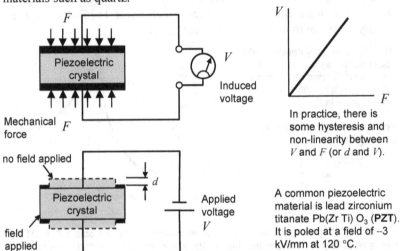

In practice, there is some hysteresis and non-linearity between V and F (or d and V).

A common piezoelectric material is lead zirconium titanate $Pb(Zr\ Ti)\ O_3$ (**PZT**). It is poled at a field of –3 kV/mm at 120 °C.

A ferroelectric material is **poled** by the application of a high voltage, thus causing the domains to become aligned. Poling at high temperature (but below T_c) allows the domain walls to move more easily.

The piezoelectric effect arises due to a change in the off-centre displacement of ions in the crystal as a stress is applied.

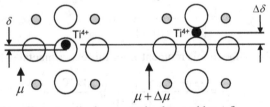

When stress is applied, the off-centre displacement is changed by $\Delta\delta$ and the dipole moment is also changed by $\Delta\mu$. The polarisation changes from P_s to P, which results in an electric field E being induced. In quartz, a non-ferroelectric material, a mechanical displacement also produces an off-centre change in the position of ions resulting in a field E being produced.

2.3.8 Piezoelectric Coupling Factor

Piezoelectric coupling factor is a measure of the efficiency for converting mechanical energy into electrical energy (and vice versa).

$$k^2 = \frac{U_e}{U_m}$$

electrical energy output

mechanical energy input

or

mechanical energy output

$$k^2 = \frac{U_m}{U_e}$$

electrical energy input

Application of strain s to a piezoelectric material results in mechanical strain energy input U_m:

Mechanical strain energy input/unit volume

$$U_m = \frac{\sigma s}{2} = \frac{1}{2}Ys^2 \quad \text{where} \quad Y = \sigma/s$$

This energy is stored as electrical potential energy U_e:

Electrical potential energy input/unit volume

$$U_e = \frac{1}{2}QV = \frac{1}{2}\varepsilon_p\varepsilon_0 E^2$$

If all the energy is transferred into electrical energy, then

$$k^2 = \frac{\varepsilon_p\varepsilon_0}{Y}\left(\frac{E}{s}\right)^2$$

$k = 0.7$ PZT
$k = 0.4$ BaTiO$_3$

In the case of a disc of area A and thickness t being acted upon by a force F and being compressed x, the coupling factor can be expressed:

$$U_m = \frac{1}{2}Fx$$ Linear elasticity where the restoring force is proportional to the displacement

$$U_e = \frac{1}{2}QV$$

$$k^2 = \frac{QV}{Fx}$$

$$Y = \frac{Fd}{Ax}$$

$$k^2 = \frac{d}{CAY}\left(\frac{F}{V}\right)^2$$

Measurements of F and V can yield a value for k. C is the capacitance of the circuit (including the disc).

$F = kx$	$U = \frac{1}{2}kx^2$
$\dfrac{F}{A} = \dfrac{kx}{A}$	$k = \dfrac{YA}{l}$
$= \dfrac{kl}{A}\dfrac{x}{l}$	$U = \dfrac{1}{2}\dfrac{YA}{l}x^2$
$Y = \dfrac{kl}{A}$	$= \dfrac{1}{2}E(Al)\left(\dfrac{x}{l}\right)^2$
$\sigma = Ys$	$\dfrac{U}{V} = \dfrac{1}{2}Ys^2$

2.3.9 AC Piezoelectric Response

In a piezoelectric material there is an induced polarisation P (arising from either the imposition of a **mechanical strain** or the application of an external field) which can be expressed:

$$P = (\varepsilon_p - 1)\varepsilon_o E$$
$$= \chi_p \varepsilon_o E$$

where ε_p is the relative permittivity in the piezoelectric state.

Consider the application of an AC voltage or a piezoelectric material. At low frequencies, the energy per unit volume stored within the material comprises that which is needed to maintain the polarisation (an electrical component) and that which is required to produce mechanical vibrations (a mechanical component).

$$U_{LF} = U_s + U_m$$

$$U_s = \frac{1}{2}\varepsilon_s \varepsilon_o E^2$$

$$U_m = k^2 U_{LF}$$

$$U_{LF} = \frac{1}{2}\varepsilon_p \varepsilon_o E^2 = \frac{1}{2}\varepsilon_s \varepsilon_o E^2 + k^2 \frac{1}{2}\varepsilon_p \varepsilon_o E^2 \quad \text{where } \varepsilon_r = \varepsilon_p$$

$$\varepsilon_s = (1 - k^2)\varepsilon_p$$

At **high frequencies**, the mechanical response of the material cannot follow the field and so there is no piezoelectric effect and thus the energy per unit volume is:

$$U_{HF} = U_s = \frac{1}{2}\varepsilon_s \varepsilon_o E^2$$

In terms of capacitance or a piezoelectric disc:

$$C_{HF} = (1 - k^2)C_{LF}$$

By measuring the high and low frequency capacitance of a crystal, the coupling constant can be estimated.

2.3.10 Piezoelectric Resonance

When a field is applied to a piezoelectric, **mechanical waves** propagate throughout the material at a frequency equal to that of the applied field. A condition for mechanical resonance exists when:

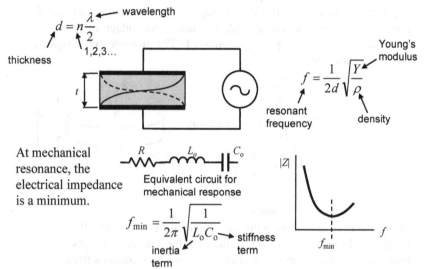

$$d = n\frac{\lambda}{2}$$

wavelength

thickness

$1,2,3...$

$$f = \frac{1}{2d}\sqrt{\frac{Y}{\rho}}$$

Young's modulus

resonant frequency

density

At mechanical resonance, the electrical impedance is a minimum.

Equivalent circuit for mechanical response

$$f_{min} = \frac{1}{2\pi}\sqrt{\frac{1}{L_o C_o}}$$

inertia term

stiffness term

$|Z|$

f_{min}

At frequencies higher than mechanical resonance, the piezoelectric effect cannot appear and so the material behaves like an ordinary dielectric in a capacitor. The equivalent circuit at high frequencies is thus:

$$|Z| = \frac{1}{\omega C}$$

Between mechanical resonance and the higher frequency regime, we have a combined mechanical/electrical system which can be represented:

At low frequencies $C_{LF} = C_o + C_1$

$$= (1 - k^2)C_{HF}$$

At high frequencies $C_{HF} = C_1$

where there is a resonance maximum at:

$$f_{max} = \frac{1}{2\pi}\sqrt{\frac{C_o + C_1}{L_o C_o C_1}}$$

The mechanical coupling factor is found from:

$$k = \sqrt{1 - \left(\frac{f_{min}}{f_{max}}\right)^2}$$

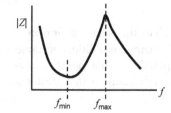

$|Z|$

f_{min} f_{max} f

2.3.11 Coupling Factor for PZT

An important piezoelectric material is Pb(Zr Ti) O_3, or PZT. While the electromechanical coupling constant for $BaTiO_3$ is about $k = 0.4$, values of k for PZT are in the range 0.5 to 0.7.

Experimentally, a series of masses is applied to a disc of material and the voltage across the disc is measured. In practice, the voltage appears only while the mass is being applied or removed. The **piezoelectric coupling factor** is found from:

$$k^2 = \frac{QV}{Fx}$$

$$Y = \frac{Fd}{Ax} \quad \text{Elastic modulus}$$

$$V = \underbrace{\left[\frac{1}{k} \sqrt{\frac{d}{CAY}} \right]}_{\text{slope}} F$$

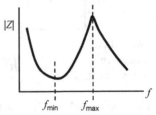

Another way of determining the coupling factor is to consider the frequency dependence on the impedance of the specimen. At low frequencies, electric dipole movement contains contributions from both the AC field and mechanical vibrations. At high frequencies, mechanical vibrations cannot follow the changes in the field due to inertia and so the induced dipole moment arises solely from the field. That is, the permittivities at high and low frequencies are related by:

$$\varepsilon'_{high} = \left(1 - k^2 \right) \varepsilon'_{low}$$

In practice, this is done by measuring the high and low frequency capacitances and impedances near a mechanical resonant condition.

$$k = \sqrt{1 - \left(\frac{f_{min}}{f_{max}} \right)^2}$$

Both methods have their own advantages: the DC method is a little inaccurate, but quite repeatable, while the AC method gives a more accurate result, with a greater scatter in the readings.

2.4 Dielectric Breakdown

Summary

$$\sigma = F \frac{n(Ze)^2 a^2 \nu}{kT} e^{-W/kT} \qquad \text{Dielectric conductivity}$$

$$\sigma \approx \frac{A}{T} e^{-W/kT}$$

$$E_{\text{breakdown}} \propto \frac{1}{\sqrt{d}} \qquad \text{Dielectric breakdown}$$

2.4.1 Dielectric Conduction

No dielectric is a perfect insulator and in some instances, the finite **conductivity** of a dielectric must be considered. Conductivity in a dielectric usually arises due to the presence of **impurities**.

The conductivity of a dielectric can manifest itself as a loss of energy which is dissipated by **resistive heating**. This is in addition to heat arising from the friction associated with dipole movement.

$$\varepsilon'' = \frac{\sigma}{\omega\varepsilon_o} + \frac{(\varepsilon_s - \varepsilon_\infty)}{1+\omega^2\tau^2}\omega\tau$$

resistive   frictional
losses   losses

In water, where the molecules have a large permanent dipole moment, impurities are easily ionised by the field associated with the dipole. Ions are then free to move throughout the liquid if an external field is applied.

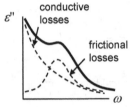

In a solid dielectric, conduction can occur when ions within the crystal structure have sufficient thermal energy to overcome the energy barrier W between adjacent atomic positions in the structure.

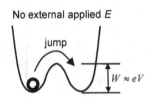

No external applied E

jump

$W \approx eV$

When no external field is applied, these jumps occur at random, but in the presence of a field, there is a preferential direction to migration of ions since the energy barrier in the direction of the field is lowered, and so a current flows.

For a three dimensional lattice, the conductivity is:

For an ion to jump, there must be an available vacant site in an adjacent position. The vacancy concentration can depend upon the concentration of impurities and also the temperature. The number of migration ions per unit volume is thus a measure of the vacancy concentration.

charge on the ion

No. migration ions per unit volume

Jump distance

frequency of vibration of ion

$$\sigma = F\frac{n(Ze)^2 a^2 v}{kT}e^{-W/kT}$$

lattice factor (e.g., = 4 for NaCl)

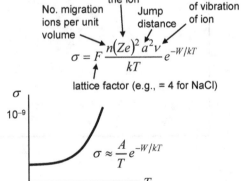

$$\sigma \approx \frac{A}{T}e^{-W/kT}$$

2.4.2 Dielectric Breakdown

Dielectric breakdown is an increase in **conductivity** of a dielectric. This may occur suddenly or gradually (from μsec to days). The process is irreversible in solids.

The field strength $E_{breakdown}$ which causes breakdown depends upon:
- Nature and composition of the dielectric
- Size and shape of the material (sharp corners)
- Environmental conditions (e.g., presence of moisture)
- Method of application of the field (AC, DC, pulsed).

The application of a sufficiently high field may cause **ionisation** of atoms within the material whereby free electrons are produced and are accelerated. This results in collisions with other atoms, which then also become ionised, and eventually there is enough ionisation and free electrons present to constitute an **electric current**.

In a gas, breakdown is often accompanied by a **discharge** of light and sound. All gases have a few ions and free electrons present as a result of bombardment by cosmic rays. Upon application of a high voltage, the free electrons initially present are accelerated and may acquire sufficient energy to ionise neutral molecules or atoms with which they collide. The discharge is maintained by ionisations from collisions.

Discharges can be classified as:
- Dark
- Glow ⟶ Applicable to low pressure gases
- Brush
- Spark

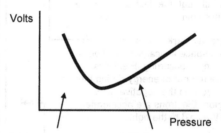

For a fixed separation of electrodes, the voltage or potential required to maintain a discharge decreases rapidly as the pressure is decreased from atmospheric. At lower pressures, the required voltage again increases.

Low pressure

Chances of collision are very much reduced due to low number density of molecules. One or two ionisations per collision.

High pressure

The presence of many gas molecules reduces the mean free path. Electrons require a greater acceleration across this shorter distance to attain sufficient energy to cause ionisation. Greater acceleration = larger potential = high voltage required.

2.4.3 Discharge Tube

In a **discharge tube**, the discharge is maintained by the emergence of electrons from the cathode as it is bombarded with +ve ions from the gas.

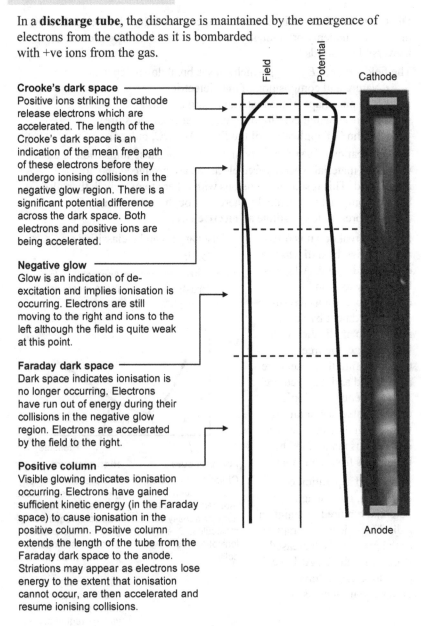

Crooke's dark space
Positive ions striking the cathode release electrons which are accelerated. The length of the Crooke's dark space is an indication of the mean free path of these electrons before they undergo ionising collisions in the negative glow region. There is a significant potential difference across the dark space. Both electrons and positive ions are being accelerated.

Negative glow
Glow is an indication of de-excitation and implies ionisation is occurring. Electrons are still moving to the right and ions to the left although the field is quite weak at this point.

Faraday dark space
Dark space indicates ionisation is no longer occurring. Electrons have run out of energy during their collisions in the negative glow region. Electrons are accelerated by the field to the right.

Positive column
Visible glowing indicates ionisation occurring. Electrons have gained sufficient kinetic energy (in the Faraday space) to cause ionisation in the positive column. Positive column extends the length of the tube from the Faraday dark space to the anode. Striations may appear as electrons lose energy to the extent that ionisation cannot occur, are then accelerated and resume ionising collisions.

2.4.4 Thermal Breakdown

Thermal breakdown occurs when heat generated by the field cannot dissipate effectively and the **temperature** of the material rises. A rise in temperature causes the conductivity of the dielectric to increase, eventually to the point where breakdown occurs.

An important parameter which characterises thermal breakdown is the ratio:

$$\frac{\sigma}{k}$$

where σ is the electrical conductivity and k is the thermal conductivity.

A large value results in a low value of $E_{\text{breakdown}}$.

Since the losses associated with an AC field include both conduction and frictional losses in the dielectric, $E_{\text{breakdown}}$ for an AC field is usually less than $E_{\text{breakdown}}$ for a DC field when breakdown is thermally activated.

The conditions for thermal breakdown can be described by comparing the rates of change of energies involved. Consider the application of a voltage across a dielectric.

$$P_E = VI$$

where P_E is the electrical power input, and

$$I = A\sigma(T)\frac{V}{L}$$

is the conduction current.

The heat lost is a function of the temperature difference between the dielectric and the surroundings, as well as the surface area exposed to the surroundings, and the thermal conductivity of the dielectric. In general, the heat lost per unit time is:

$$\sigma(T) = Ae^{-W/kT}$$

$$\approx \sigma_0 \exp\left(\frac{W}{k}(T - T_0)\right)$$

$$\dot{Q} = AC(T - T_0)$$

where A is the surface area and C is a constant.

Equilibrium is established when the power input becomes equal to the heat lost per unit time. When the power input exceeds the heat lost, then the temperature of the dielectric rises until breakdown occurs.

Characteristics of thermal breakdown:

- There may be some time before the onset of breakdown while the dielectric heats up.

- $E_{\text{breakdown}} \propto \dfrac{1}{\sqrt{d}}$ for thin samples, where d is the thickness.

2.5 Examples of Dielectrics

Summary

$$\varepsilon_{r\,(\text{solid})} > \varepsilon_{r\,(\text{liquid})} > \varepsilon_{r\,(\text{gas})} \qquad \text{Relative permittivities}$$

$$N = N_0 \frac{\rho}{m} \qquad \text{Dipole density}$$

$$N\alpha = N_1\alpha_1 + N_2\alpha_2 + ... \qquad \text{Mixed dipoles}$$

$$\alpha = \alpha_1 + \alpha_2 + ... \qquad \text{Multiple polarisabilities}$$

Kirkwood equation

$$\frac{(\varepsilon_r - 1)(2\varepsilon_r + 1)}{9\varepsilon_r} = \frac{N}{3\varepsilon_o}\left(\alpha_e + g\frac{\mu^2}{3kT}\right)$$

$$\frac{\varepsilon_r - 1}{\varepsilon_r + 2} = \frac{N\alpha_e}{3\varepsilon_o} \qquad \text{Clausius–Mosotti Equation}$$

2.5.1 Dielectric Properties

Some interesting features of the dielectric properties of materials.

1. $\varepsilon_r \to \infty$ as $\dfrac{N\alpha}{3\varepsilon_0} \to 1$

2. ε_r increases with increasing N so that:

$\varepsilon_{r \text{ (solid)}} > \varepsilon_{r \text{ (liquid)}} > \varepsilon_{r \text{ (gas)}}$

3. α only depends on the nature of the atomic dipole and not on the nature of the environment.

4. N can be expressed in terms of Avogadro's number $N_0 = 6.023 \times 10^{23}$ mol^{-1}

$$N = N_0 \frac{\rho}{m}$$

density kg/m³

mass kg

5. At optical frequencies $\varepsilon_r = n^2$ where n is the refractive index of the material.

6. When the material contains a number of different dipole types, then $N\alpha$ is given by the sum of the contribution from each:

$$N\alpha = N_1\alpha_1 + N_2\alpha_2 + ...$$

No. per unit
volume of
type 1

7. When an atom or a molecule contributes in more than one way to the overall dipole moment μ_E then the overall polarisibility α is given by:

$$\alpha = \alpha_1 + \alpha_2 + ...$$

2.5.2 Polarisability

Consider an oxide ceramic with a relative permittivity of 6.8 over a range of frequencies from microwave to infrared. At optical frequencies, the refractive index is 1.48. The structure is a cubic unit cell of 5.5×10^{-10} m.

At optical frequencies,

$$\varepsilon_r = n^2$$
$$= 1.48^2$$
$$= 2.19$$

$$\frac{\varepsilon_r - 1}{\varepsilon_r + 2} = \frac{N\alpha_e}{3\varepsilon_0}$$

$$\frac{1.19}{4.19} = \frac{4(\alpha_e)}{3(8.85 \times 10^{-12})(5.5 \times 10^{-10})^3}$$

$$\alpha_e = 3.13 \times 10^{-40} \, \text{F m}^2 \quad \text{per molecule}$$

Below infrared frequencies:

$$\varepsilon_r = 6.8$$

$$\frac{\varepsilon_r - 1}{\varepsilon_r + 2} = \frac{N\alpha_e}{3\varepsilon_0}$$

$$\frac{5.8}{8.8} = \frac{4(\alpha_i + \alpha_e)}{3(8.85 \times 10^{-12})(5.5 \times 10^{-10})^3}$$

$$= \frac{4\alpha_i}{3(8.85 \times 10^{-12})(5.5 \times 10^{-10})^3} + 0.284$$

$$\alpha_i = 4.14 \times 10^{-40} \, \text{F m}^2 \quad \text{per molecule}$$

2.5.3 Dielectric Properties of Water

Water presents an extremely interesting case of dielectric loss because of the strong molecular permanent dipole moment.

The **Clausius–Mosotti equation** expresses the relative permittivity in terms of the atomic dipole moment μ and the polarisability α for polar molecules where the local field is different from the applied external field due to the proximity of nearby molecules.

$104°$

$\mu = 1.9$ debyes

$$\frac{\varepsilon_r - 1}{\varepsilon_r + 2} = \frac{N\alpha}{3\varepsilon_0}$$

electronic → dipolar

$$= \frac{N}{3\varepsilon_0}\left(\alpha_e + \frac{\mu^2}{3kT}\right)$$

$$= \frac{N}{3\varepsilon_0}\alpha_e + \left(\frac{N}{3\varepsilon_0}\frac{\mu^2}{3k}\right)\frac{1}{T}$$

intercept term   slope term

N is the number of dipoles per unit volume

$$N = \frac{N_o\rho}{m.w.}$$

$$= \frac{6.023 \times 10^{23}(1000)}{18 \times 10^{-3}}$$

$$= 3.34 \times 10^{28}$$

The Clausius–Mosotti equation is best when the dipole moments are small and well-separated. For water, this is not the case, and the **Kirkwood equation** is found to be a more accurate description of the relative permittivity.

$$\frac{(\varepsilon_r - 1)(2\varepsilon_r + 1)}{9\varepsilon_r} = \frac{N}{3\varepsilon_0}\left(\alpha_e + g\frac{\mu^2}{3kT}\right)$$

correlation parameter = 2.68 for H_2O

Plotting $\dfrac{(\varepsilon_r - 1)(2\varepsilon_r + 1)}{9\varepsilon_r}$ vs $\dfrac{1}{T}$ gives

slope term $\dfrac{N}{3\varepsilon_0}\dfrac{\mu^2}{3k}g$

intercept term $\dfrac{N\alpha_e}{3\varepsilon_0}$

Can be measured experimentally to obtain values of μ and α_i by measuring the capacitance of a parallel plate capacitor with a water dielectric over a range of temperatures.

Typical values from an experiment are $\mu = 0.226$ debye for C-M, and $\mu = 2.5$ debye for Kirkwood.

Experiments show that ε_r decreases with increasing temperature as the external field finds it harder to align dipoles under increasing thermal agitation.

2.5.4 Dielectric Properties of Paper

Paper is an important dielectric material because its ready availability, low cost, and mechanical and thermal properties make it ideal for use in capacitors.

It consists of a network of fibres and pockets either filled with air, water or oil depending on its treatment. To a first order approximation, the equivalent circuit consists of two capacitors C_1 and C_2 in series with leakage terms R_1 and R_2. Thus:

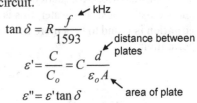

Individual fibres | spaces between the fibres

$$\varepsilon' = \varepsilon_\infty + \frac{\varepsilon_s - \varepsilon_\infty}{1 + (\omega\tau)^2} \quad \begin{array}{l}\text{Energy stored}\\ \text{in dielectric}\end{array} \qquad \varepsilon'' = \frac{\sigma}{\omega\varepsilon_0} + \frac{(\varepsilon_s - \varepsilon_\infty)\omega\tau}{1 + (\omega\tau)^2} \quad \begin{array}{l}\text{Energy}\\ \text{dissipated}\end{array}$$

In terms of the equivalent circuit,

$$\varepsilon_s = \frac{R_1^2 C_1 + R_2^2 C_2}{C_0 (R_1 + R_2)^2} \qquad \varepsilon_\infty = \frac{C_1 C_2}{C_0 (C_1 + C_2)} \qquad \tau = \frac{C_1 + C_2}{1/R_1 + 1/R_2}$$

DC relative permittivity | Relative permittivity $f = \infty$ | Effective time constant

If testing is done over a reasonably high frequency range (>10 kHz) then the conductivity term σ is small and so

Plotting ε''/ω vs ε' can be done by measuring R, C and $\tan\delta$ as a function of frequency using a bridge circuit.

$$\frac{\varepsilon''}{\omega} \approx \varepsilon'\tau - (\varepsilon''\tau)$$

slope   intercept

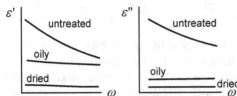

$$\tan\delta = R\frac{f}{1593}$$

distance between plates

$$\varepsilon' = \frac{C}{C_0} = C\frac{d}{\varepsilon_0 A}$$

area of plate

If untreated, dried and oily paper are tested, experiments show that:

$$\varepsilon'' = \varepsilon'\tan\delta$$

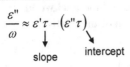

	τ sec	ε_∞
dried	1.6×10^{-5}	1.9
oily	9.8×10^{-6}	2.6
untreated	2.03×10^{-5}	3.1

These results indicate that untreated paper has a high loss ε'' and marked frequency dependence consistent with the effect of dipolar water molecules within it, while oily paper has low loss with a stable value of ε' over the frequency range, making it suitable for use as a dielectric in a **capacitor**.

2.5.5 Transformer Oil

In this example, the real and imaginary parts of the relative permittivity are calculated along with the loss factor at a frequency of 1.5915 kHz where a **transformer oil** is used as the dielectric between the plates of a parallel plate **capacitor** of diameter 100 mm and a separation of 0.2 mm. The resistivity of the oil has been measured at 2.5×10^{10} Ω m and the capacitance has been measured at 950 pF.

$$
\left.
\begin{aligned}
A &= \pi(0.05)^2 \\
&= 7.85 \times 10^{-3} \, \text{m}^2 \\
C_o &= \varepsilon_o \frac{A}{d} \\
&= \frac{8.85 \times 10^{-12} (7.85 \times 10^{-3})}{0.0002} \\
&= 3.47 \times 10^{-10} \, \text{F}
\end{aligned}
\right\} \begin{aligned} &\text{From dimensions} \\ &\text{given} \end{aligned}
$$

$$
\boxed{\begin{aligned} C &= \varepsilon_r \left[\varepsilon_o \frac{A}{d} \right] \\ C_o &= \varepsilon_o \frac{A}{d} \end{aligned}}
$$

$$
\varepsilon' = \frac{C}{C_o} \qquad\qquad \varepsilon'' = \frac{\sigma}{\omega \varepsilon_o}
$$

$$
= \frac{950 \times 10^{-12}}{3.47 \times 10^{-10}} \qquad = \frac{1}{2.5 \times 10^{10}} \frac{1}{2\pi(1.5915 \times 1000)(8.85 \times 10^{-12})}
$$

$$
= 2.73 \qquad\qquad = 4.52 \times 10^{-4}
$$

$$
\tan \delta = \frac{\varepsilon''}{\varepsilon'}
$$

$$
= \frac{4.52 \times 10^{-4}}{2.73}
$$

$$
= 1.6 \times 10^{-4}
$$

If 10 V_{rms} is applied, the power dissipation is:

$$
P_{av} = \frac{V_{rms}^2}{R} = V_{rms}^2 \sigma \frac{A}{d}
$$

$$
= V_{rms}^2 \varepsilon'' \omega C_o
$$

$$
= 10^2 (4.52 \times 10^{-4}) 2\pi(1591.5)(3.47 \times 10^{-10})
$$

$$
= 0.156 \, \text{W}
$$

2.5.6 Contaminated Transformer Oil

In this example, the relative permittivity of a **transformer oil** contaminated with water is measured to be 4.25. The relative permittivity of uncontaminated water is 3.40 and the relative permittivity of water is 70.0. From this data, it is possible to calculate the volume fraction of absorbed water.

The **Clausius–Mosotti equation** is used to determine the polarisability of water and oil separately.

$$\frac{\varepsilon_r - 1}{\varepsilon_r + 2} = \frac{N\alpha}{3\varepsilon_0}$$

$$\frac{69}{72} = \frac{N_w \alpha_w}{3(8.85 \times 10^{-12})} \quad \text{water}$$

$$N_w \alpha_w = 25.4 \times 10^{-12}$$

$$\frac{2.4}{5.4} = \frac{N_o \alpha_o}{3(8.85 \times 10^{-12})} \quad \text{oil}$$

$$N_o \alpha_o = 11.8 \times 10^{-12}$$

$$N_w \alpha_w + N_o \alpha_o = f_w 25.4 \times 10^{-12} + f_o 11.8 \times 10^{-12}$$

The Clausius–Mosotti equation is then applied to the mixture.

fraction water   fraction oil $f_w + f_o = 1$

$$\frac{4.25 - 1}{4.25 + 2} = \frac{f_w 25.4 \times 10^{-12} + f_o 11.8 \times 10^{-12}}{3(8.85 \times 10^{-12})}$$

$$13.8 = f_w 25.4 + f_w 11.8$$

$$= (1 - f_o)15.4 + f_o 11.8$$

$$f_o = 85.3\%$$

$$f_w = 14.7\%$$

If, at 50 Hz, the loss factor is found to be 0.19, then the conductivity of the oil is calculated from:

$$\tan \delta = 0.19 = \frac{\varepsilon''}{\varepsilon'}$$

$$\varepsilon'' = 4.35(0.19)$$

$$= \frac{\sigma}{\omega \varepsilon_0}$$

$$\sigma = 50(2\pi)(8.85 \times 10 - 12)(4.25)(0.19)$$

$$= 2.24 \times 10^{-9} \text{S} \quad \text{(siemens)}$$

2.5.7 Sodium Chloride

The data pertains to NaCl. Using this data, we can calculate the **relative permittivity** at DC conditions.

Diameter of Na+ ion	1.8×10^{-10} m
Diameter of CL- ion	2.4×10^{-10} m
Density of NaCl	2163 kg m^{-3}
a.w. Na	23
a.w. Cl	35.5
Lattice frequency	5×10^{12} Hz

$$Q_e = 0.7e$$

$$\omega_o = 2\pi \left(5 \times 10^{12}\right)$$

$$\frac{\varepsilon_r - 1}{\varepsilon_r + 2} = \frac{N}{3\varepsilon_o}\left(\alpha_e + \alpha_i\right) \quad \text{at DC}$$

$$\alpha_e = 4\pi\varepsilon_o R^3$$

$$= 4\pi\varepsilon_o \left(0.9 \times 10^{-3^3} + 1.2 \times 10^{-10^3}\right)$$

$$= 2.73 \times 10^{-40}\,\text{Fm}^2$$

$$m_1 = \frac{0.023}{N_o}$$

$$m_2 = \frac{0.0355}{N_o}$$

$$\frac{1}{m} = \frac{1}{m_1} + \frac{1}{m_2}$$

$$\alpha_i = \frac{Q^2}{m\omega_o^2}$$

$$= \frac{0.7\left(1.6 \times 10^{-19}\right)^2 N_o}{2\pi\left(5 \times 10^{12}\right)}\left(\frac{1}{0.023} + \frac{1}{0.0355}\right)$$

$$= 5.48 \times 10^{-40}\,\text{F m}^2$$

$$\frac{\varepsilon_r - 1}{\varepsilon_r + 2} = \frac{N_o \rho}{3\varepsilon_o \omega}\left(5.48 \times 10^{-40} + 2.73 \times 10^{-40}\right)$$

$$\varepsilon_r = 7.65$$

At optical frequencies, $\alpha = \alpha_e$ and so:

$$\frac{\varepsilon_r - 1}{\varepsilon_r + 2} = \frac{N_o \rho}{3\varepsilon_o \omega} 2.73 \times 10^{-40} = n^2$$

$$n = 1.37$$

2.5.8 Ferroelectric Ceramic

The domain wall movement in a **ferroelectric** material is very small when an external field is applied.

Consider a ceramic disc, 1 mm thick, with a relative permittivity of 8450 and a spontaneous polarisation of 0.32 Cm^{-2}. If the average domain size is 0.85 μm, then, when a 100 V peak to peak voltage is applied, the domain wall movement can be calculated from:

$$\varepsilon_r = \frac{2P_s}{D\varepsilon_o}\frac{dx}{dE}$$

$$8450 = \frac{2(0.32)}{0.85\times10^{-6}\left(8.85\times10^{-12}\right)}\frac{dx}{dE}$$

$$\frac{dx}{dE} = 9.93\times10^{-14}\,\text{mV}^{-1}$$

$$\Delta E = \frac{\Delta V}{d}$$

$$= \frac{100}{1\times10^{-3}}$$

$$= 100000\ \text{Vm}^{-1}$$

$$dx = 100000\left(9.93\times10^{-14}\right)$$

$$= 9.9\times10^{-9}\,\text{m}\ \text{ total movement from }-50\text{ V to }+50\text{ V}$$

$$= 5\times10^{-9}\,\text{m}\ \longrightarrow\ \text{From equilibrium position}$$

2.5.9 Ferroelectric Permittivity

Practical measurements of the relative permittivity of ferroelectric materials are influenced by the presence of small air pockets between the electrodes and the specimen. The measured capacitance is thus usually less than expected since there are now essentially two capacitors in series to be considered.

$$\dashv\vdash\dashv\vdash$$
$$C_{air} \quad C$$

Consider a disc shaped ferroelectric material of thickness $d = 2$ mm and diameter 20 mm with $\varepsilon_r = 8400$ where the measured capacitance, $C_m = 5.561$ nF. The expected capacitance can be determined from the dimensions of the specimen:

$$C = \varepsilon_r \varepsilon_o \frac{A}{d}$$

$$= 8400 \big(8.85 \times 10^{-12} \big) \frac{\pi (0.01)^2}{0.002}$$

$$= 11.7 \text{ nF}$$

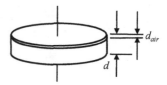

where it is assumed that the thickness of the air gap is very much smaller than the overall thickness of the specimen.

Taking into consideration the average effect of the multitude of air pockets, we have:

$$\frac{1}{C_m} = \frac{1}{C_{air}} + \frac{1}{C}$$

$$\frac{1}{5.561 \times 10^{-9}} = \frac{1}{C_{air}} + \frac{1}{1.17 \times 10^{-8}}$$

$$C_{air} = 1.06 \times 10^{-8} \text{ F}$$

The effective or average thickness of the air gap is thus:

$$d_{air} = \frac{\varepsilon_o A}{1.06 \times 10^{-8}}$$

$$= \frac{\big(8.85 \times 10^{-12} \big) \pi (0.01)^2}{1.05 \times 10^{-8}}$$

$$= 2.62 \times 10^{-7} \text{ m}$$

Part 3

Magnetic Properties of Materials

3.1 Magnetic Field

Summary

$$F = qv \times B$$ Magnetic force

$$B = \frac{\Phi}{A}$$ Magnetic flux density

$$D = \varepsilon E$$ Electric flux density

$$H = \frac{B}{\mu}$$ Magnetic field

$$u = \frac{1}{2}\frac{B^2}{\mu_o} = \frac{1}{2}\mu_o H^2$$ Energy density

3.1.1 Magnetic Field

Over and above any electrostatic force

A **magnetic field** B is said to exist at a point if a force is exerted on a moving charge at that point. The force acting on a **moving charge** is perpendicular to both the direction of the field and the velocity of the charge.

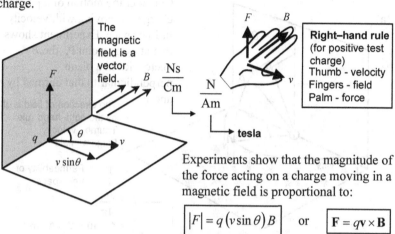

The magnetic field is a vector field.

$B \dfrac{Ns}{Cm}$ $\dfrac{N}{Am}$ → **tesla**

Right–hand rule (for positive test charge)
Thumb - velocity
Fingers - field
Palm - force

Experiments show that the magnitude of the force acting on a charge moving in a magnetic field is proportional to:

$$|F| = q(v\sin\theta)B \quad \text{or} \quad \mathbf{F} = q\mathbf{v}\times\mathbf{B}$$

B is called the **magnetic induction** or the **magnetic field**.

A magnetic field may be represented by lines of **induction**. The **magnetic flux** is proportional to the total number of lines.

Note: not **lines of force** since, unlike the electric field, the magnetic force is perpendicular to the direction of the field.

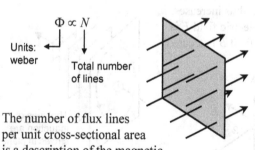

$\Phi \propto N$

Units: weber

Total number of lines

Uniform magnetic field - lines are equally spaced.

The number of flux lines per unit cross-sectional area is a description of the magnetic field and is called the **magnetic induction** B or **magnetic flux density** (tesla).

Spacing between field lines indicates field strength.

$$B = \frac{\Phi}{A}$$

1 tesla = 1 weber per square metre

3.1.2 Source of Magnetic Fields

A **magnetic field** can be created in two ways: (a) by the movement of charge carriers in a conductor (i.e., an electric current) and (b) by a changing **electric field** in an insulator (or empty space).

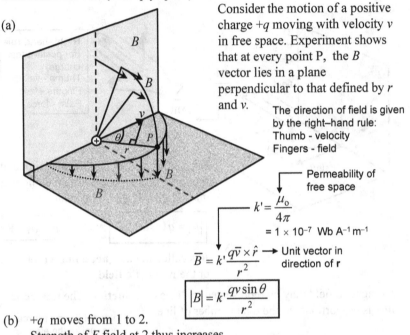

(a)

Consider the motion of a positive charge $+q$ moving with velocity v in free space. Experiment shows that at every point P, the B vector lies in a plane perpendicular to that defined by r and v.

The direction of field is given by the right–hand rule:
Thumb - velocity
Fingers - field

Permeability of free space

$$k' = \frac{\mu_o}{4\pi}$$
$$= 1 \times 10^{-7} \text{ Wb A}^{-1}\text{m}^{-1}$$

$$\overline{B} = k'\frac{q\overline{v} \times \hat{r}}{r^2}$$ → Unit vector in direction of **r**

$$|B| = k'\frac{qv\sin\theta}{r^2}$$

(b) $+q$ moves from 1 to 2.
Strength of E field at 2 thus increases.
In a space where the electric field E is changing, a magnetic field B is created.

The electric field lines radiate outwards from the charge.

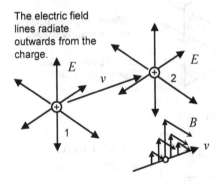

The lines of induction are circles in planes perpendicular to the velocity.

Vertical components of E along the line of v are shown. At that point, the B field is horizontal.

3.1.3 Magnetic Flux Density

A magnetic field may be represented by lines of what we call magnetic induction. The total number of lines passing through a cross-sectional area is called the **magnetic flux**. The number of lines per unit cross-sectional area is the **magnetic flux density** or more simply, **magnetic field**, and is given the symbol B. B is analogous to D for the electric field case – however, unlike the electric field, where the force on a charged particle depends upon E, the **electric field intensity**, the force on a moving charged particle in a magnetic field depends upon B, the magnetic flux density.

Analogous to E for the electric case, we have the quantity H to represent the **magnetic field intensity** or **magnetic field strength** for magnetic fields. It seems odd that the magnetic force experienced by a moving charge should depend on B and not H, but as we shall see, this is a consequence of the way in which materials behave in a magnetic field.

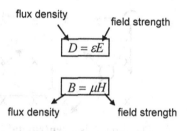

flux density field strength
$$D = \varepsilon E$$
$$B = \mu H$$
flux density field strength

In the case of an electric field, D is applied (by the free charges) and E is the result – the magnitude of which depends upon ε. In the magnetic case, *because it is B that determines the magnetic force*, it is H that is applied to a medium or a material and B is the result. The ease with which a magnetic field B is set up within a material depends upon the material property called the **permeability** μ of the material. In this case, we think of the magnetic field intensity H as being modified by an "effectiveness factor" μ to produce a magnetic flux density B. In the electrical case, we considered the electric flux density D to be modified by an effectiveness factor $1/\varepsilon$ to produce lines of field intensity E. Our interest is primarily in the physical quantities, E and B, that lead to forces on charged particles since it is these forces that are responsible for physical phenomena.

In the case of an electric field between two parallel plates, for a constant value of flux density D arising from the free charges on the plate, when a dielectric was inserted, the field strength E was reduced due to a reverse polarisation field from the dielectric. We shall see that in the case of a magnetic field, say in a coil, when a magnetic material is present, the field strength H for a constant current I remains unchanged, but the flux density B changes (can be decreased but usually increased) due to magnetisation within the material. For a given cross-sectional area A, if B increases, then so does the flux.

3.1.4 Charged Particle in Magnetic Field

1. A positively charged particle is
   given velocity v in a direction
   perpendicular to a uniform magnetic
   field B.

2. A force $F = qvB$ is
   exerted on the
   particle downwards.

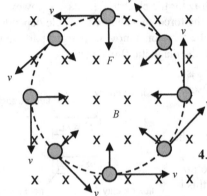

3. Application of force
   changes the direction of
   motion of the particle.

4. If the motion of the particle is
   completely within the field,
   then the particle travels in a
   circle of radius R with
   constant **tangential speed** $|v|$.

5. The force F is a **centripetal force**:

$$F = qvB$$

$$= m\frac{v^2}{R}$$

$$R = \frac{mv}{Bq}$$
   radius of path

$$\omega = \frac{qB}{m}$$
   frequency

If the direction of motion is not
perpendicular to the field, then
the velocity component parallel
to the field remains constant and
the particle moves in a helix.

Magnetic poles

It is a peculiar property of magnetic field lines that they always form closed loops.
Electric field lines may start on an isolated positive charge and terminate on another
isolated negative charge. Magnetic field lines do not start and finish on isolated
magnetic poles even though we may draw them as starting from the north pole of a
magnet and finishing on the south pole. Magnetic field lines actually pass through
the magnet to join up again.

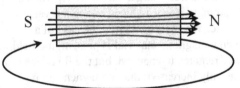

Mathematically, this is
equivalent to saying that:

$$\oint \mathbf{B} \bullet d\mathbf{A} = 0$$

Gauss' law for magnetism.

3.1.5 Force on a Current-Carrying Conductor

Consider the movement of both positive and negative charge carriers in a conductor perpendicular to a magnetic field B.

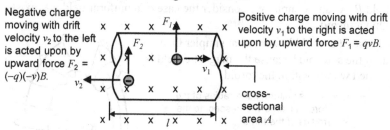

Negative charge moving with drift velocity v_2 to the left is acted upon by upward force $F_2 = (-q)(-v)B$.

Positive charge moving with drift velocity v_1 to the right is acted upon by upward force $F_1 = qvB$.

cross-sectional area A

Let n_1 and n_2 be the number of positive and negative charge carriers per unit volume. The total number of charge carriers N in a length l and cross-sectional area A of the conductor is:

$$N = n_1 Al + n_2 Al$$

$$F = n_1 Al(q_1 v_1 B) + n_2 Al(q_2 v_2 B)$$

$$= (n_1 q_1 v_1 + n_2 q_2 v_2)AlB$$

$$= JAlB$$

$$= IlB$$

$$\boxed{\mathbf{F} = I\mathbf{l} \times \mathbf{B}}$$

total force on all charge carriers both positive and negative

J current density

The cross product provides the information about the direction of F.

Note: here we have considered the movement of both positive and negative charge carriers within a **conductor**. If current flows due to the movement of only one type of charge, (e.g., electrons in a metal), then, from the macroscopic point of view, this is *exactly* equivalent to the equal movement of only positive charge carriers in the opposite direction.

The resultant force on the loop is zero.

The resultant **torque** (or moment) is:

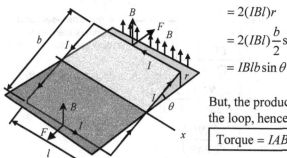

$$M_x = 2Fr$$

$$= 2(IBl)r$$

$$= 2(IBl)\frac{b}{2}\sin\theta$$

$$= IBlb\sin\theta$$

But, the product lb = the area A of the loop, hence:

$$\boxed{\text{Torque} = IAB\sin\theta}$$

The product IA is called the **magnetic moment** of the loop and is (unfortunately) given the symbol μ.

3.1.6 Energy in a Magnetic Field

A magnetic field has the potential to do work. It is of interest therefore to determine what potential **magnetostatic energy** is available at any point in a field B. As an example, we consider the case of a uniform field that exists within a **toroid**.

The case of a toroid is somewhat simpler than that of the solenoid because the magnetic field is confined wholly within the toroid.

> └──→ and is uniform if the radius of the toroid is large with respect to the radius of the turns.

The inductance of an air-filled toroid is given by:

$$L = \mu_o \frac{N^2 A}{l}$$

The length $l = 2\pi R$ is the circumference of the toroid and the area A is the cross-sectional area of the loops.

The use of μ_o here signifies the inductance of a toroid with an air (or strictly speaking, vacuum) core. The inductance of a toroid may be significantly increased when the coil is wound on a material with a high **permeability**.

An inductor stores energy in its magnetic field (much like a capacitor stores potential energy in its electric field). To determine the energy stored in an inductor, we start with the expression for the voltage across an inductor (Faraday's law):

$$V = L\frac{di}{dt}$$

$$P = VI$$

$$U = \int P dt$$

$$= \int_0^t LI\frac{dI}{dt} dt$$

$$= L\int_0^I I dI$$

$$= \frac{1}{2} LI^2$$

For the uniform field within a toroid, the energy density u is U per unit volume:

$$u = \frac{U}{V}$$

$$= \frac{1}{2}\frac{LI^2}{A2\pi R}$$

V is the cross–sectional area times the mean circumference.

$$= \frac{1}{2}\frac{\mu_o N^2 A}{2\pi R}\frac{I^2}{A2\pi R}$$

inductance for an air-filled toroid: $L = \mu_o \dfrac{N^2 A}{2\pi R}$

$$= \frac{1}{2}\mu_o \left(\frac{NI}{2\pi R}\right)^2$$

$$\boxed{u = \frac{1}{2}\frac{B^2}{\mu_o} = \frac{1}{2}\mu_o H^2}$$

Energy density in a magnetic field J/m^3

3.2 Magnetic Moment

Summary

$$\boldsymbol{\mu_M} = g\left(\frac{-q_e}{2m_e}\right)\mathbf{J}$$
Magnetic moment (electron)

$$\mu_z = g\mu_B s$$
Bohr magneton

$$U = 2\left(+\frac{q_e\hbar}{2m_e}\right)Bs$$
Energy of magnetic moment

$$U = g\mu_B B j$$
Zeeman splitting

$$\omega = \frac{q_e}{2m_e}Bm$$
Larmor precession

3.2.1 Magnetic Moment

Experiments show that:

- a charge moving perpendicular to magnetic field lines experiences a magnetic force.

$$|F| = qvB \sin \phi$$

- a magnetic field is produced by a moving charge.

$$|B| = k' \frac{qv \sin \theta}{r^2}$$

These two phenomena are a consequence of the natural tendency of magnetic fields to align themselves (since this is a position of minimum potential energy). Consider the magnetic field created by the moving charges in the wire windings of a solenoid coil:

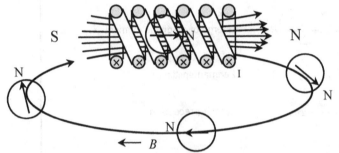

Note, the north geographic pole of the earth is a magnetic south pole. The north pole of a compass needle points in the direction of the field lines.

The ends of such a magnetised coil are commonly labelled **North** and **South** to indicate the direction of the external field. These labels have come about since it is the direction in which the solenoid would tend to align itself with the magnetic field of the earth if it were free to move.

A compass needle itself consists of a north and south pole, it being a small magnet free to rotate on a spindle. A compass needle placed in the vicinity of the coil would tend to align itself with the field surrounding the coil. Now, outside the coil, the alignment results in the familiar observation that like poles repel and unlike poles attract. Inside the coil, the compass needle is still aligned with the field but we can no longer say that like poles repel and unlike poles attract. Rather, it is more scientifically appropriate to say that *when two magnetic fields interact, they tend to align themselves.* This tendency to alignment exerts a mutual torque (or moment) between the bodies producing the fields. This is another way of expressing the concept of a **magnetic moment**.

3.2.2 Magnetic Moment of an Electron

All the magnetic properties of materials are due to the motion of electric charges within atoms. Such movement is usually due to orbiting and spinning electrons around the nucleus.

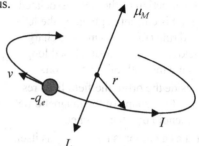

Consider a single electron in orbit around a nucleus with an angular velocity ω (ignore the spinning motion of the electron just now).

The **angular momentum** L of the orbiting electron is given by:

$$L = m_e r^2 \omega$$

The area A of the path traced out by the electron is:

$$A = \pi r^2$$

The electric current associated with the orbital motion of the electron is:

$$I = q \frac{\omega}{2\pi}$$

The **magnetic moment** is similar in a sense to the electrical dipole moment formed by an electric dipole placed in an electric field. In magnetism, there are no magnetic charges to form dipoles (which is why when a bar magnet is cut, each piece has its own N and S poles). The term **magnetic *dipole* moment** is therefore not applicable and so we say **magnetic moment** instead.

The **orbital magnetic moment** μ_M is given by the product of I and A, hence:

$$\mu_M = q \frac{\omega}{2\pi} \pi r^2$$

$$= \left(\frac{-q_e}{2m_e} \right) L$$

The physical significance of the negative sign means that μ in this case is in the opposite direction to L because for an electron, $q = -q_e = -1.6 \times 10^{-19}$ C.

The above expression applies to the orbital motion of the electron around the nucleus. It can be shown from quantum mechanics where the angular momentum arises from the spinning motion of the electron, the **spin magnetic moment** is expressed:

$$\mu_M = \left(\frac{-q_e}{m_e} \right) S$$ where S is the spin angular momentum.

The total magnetic moment for an atom has contributions from both orbital and spin motion of several electrons. In general, we have:

$$\boxed{\mu_M = g \left(\frac{-q_e}{2m_e} \right) J}$$

g is the Lande factor (equal to 1 for pure orbital motion and 2 for pure spin), and in this formula, J is the total angular momentum from the combination of spin and orbital motions for all the electrons.

3.2.3 Magnetic Field of an Orbiting Electron

An orbiting electron around the nucleus of an atom constitutes an **electric current** I and thus generates a magnetic field B_o.

The direction of the field associated with this current is given by the left-hand rule (for a moving –ve charge). Field lines for B_o form closed loops around the path of the electron.

Outside the orbit, the field lines resemble that of the shape of the magnetic field of a solenoid or a bar magnet.

Just as a compass needle aligns itself with the Earth's field, an atom with a net magnetic moment tends to align itself with the field. The response of a magnetic moment in a magnetic field is similar to that of an **electric dipole** in an electric field – but there is an important difference. In the electrical case, when the dipole aligns with the field, the total net field is reduced by the field E_d arising from polarisation of the dielectric. This is because field lines in an electric dipole begin and end on isolated charges. In the magnetic case, field lines form closed loops. When an energised solenoid, a bar magnet or an atomic magnetic moment aligns with the field, the field in the vicinity of the moment is generally increased, not reduced. This is because the portion of the field lines inside the solenoid, permanent magnet or electron orbit are closer together and more concentrated than those on the outside. If many atoms were stacked upon each other, and the stack bent around into the shape of a toroid, then the magnetic field lines B_o would be entirely contained within the toroid and lie inside the orbit.

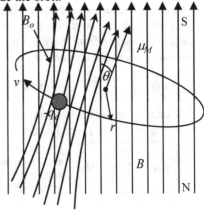

The concentrated field lines add to the applied external field and the net field is increased. This is called **paramagnetism**. However, as we shall see, another consequence of the motion of an electron around an atom is a slight reduction in the magnetic field due to the action of the **Lorentz force** on the moving electron. This is called **diamagnetism** and is a much weaker effect than paramagnetism.

3.2.4 Quantum Numbers

In an atom, at each value of the **principal quantum number** n corresponds to the allowable energy levels E for the surrounding electrons where $n = 1,2,3,...$ and $E = -\dfrac{1}{(4\pi\varepsilon_o)^2}\dfrac{m_e q_e^4}{2n^2\hbar^2}$

The orbital **angular momentum** L can take on several distinct values. Each of the values is described by a second quantum number l. The allowed values of l are 0, 1, ... $(n-1)$. Each value of l corresponds to an energy level indicated by a letter. The allowable values of the angular momentum are:
$$L = \sqrt{l(l+1)}\,\frac{h}{2\pi} = \sqrt{l(l+1)}\,\hbar$$
$$l = 0,1,2,...n-1$$

The letters have historical significance from spectroscopy experiments.

$l = 0$ s
$l = 1$ p
$l = 2$ d
$l = 3$ f
$l = 4$ g
$l = 5$ h

A third quantum number m describes the allowable changes in angle of the orbital angular momentum vector in the presence of a magnetic field. It takes the integer values $-l$ to 0 to $+l$. What this means is that the z component of L is constrained to have values in increments such that:

$$L_z = m\frac{h}{2\pi} = m\hbar$$
$$m = -l, -l+1,...l-1, l$$

Allowed values of Lz for $l = 2$

Note that L_z cannot ever equal L. The largest component of Lz is $l\hbar$.

For a given l, there are $2l + 1$ possible states or values of L_z.

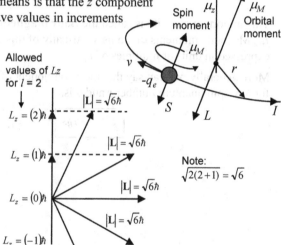

$L_z = (2)\hbar$ $|\mathbf{L}| = \sqrt{6}\hbar$

$L_z = (1)\hbar$ $|\mathbf{L}| = \sqrt{6}\hbar$

$L_z = (0)\hbar$ $|\mathbf{L}| = \sqrt{6}\hbar$

$|\mathbf{L}| = \sqrt{6}\hbar$

$L_z = (-1)\hbar$

Note:
$\sqrt{2(2+1)} = \sqrt{6}$

The fourth quantum number s describes the **spin** of an electron where the **spin quantum number** can be either $s = +1/2$ (up) or $-1/2$ (down) and indicates whether the z component of the **spin angular momentum** S_z is aligned with (up) or opposite (down) to any external magnetic field (usually oriented along the z axis). S_z can thus take on two values:

$$S_z = \pm s\hbar = \pm\frac{h}{4\pi} = \pm\frac{\hbar}{2}$$

3.2.5 Bohr Magneton

The spin magnetic moment of electrons in atoms is responsible for the most part for the magnetic properties of matter. The spin magnetic moment of a single electron is:

$$\mu_M = \left(\frac{-q_e}{m_e}\right) S$$

where the z component of S is constrained to have values:

$$S_z = \pm\frac{h}{4\pi} = \pm\frac{\hbar}{2} = \pm s\hbar$$

Thus, the z component of the spin magnetic moment associated with a single electron is expressed:

$$\mu_z = \pm\frac{1}{2}\frac{q_e}{m_e}\hbar = \pm9.3\times10^{-24}\,\text{A m}^2$$

This important quantity is called the **Bohr magneton** and given the symbol μ_B. Magnetic moments of atoms are usually of this order and are usually expressed in units or multiples of μ_B.

More generally, we can say that the z component of the magnetic moment for any spin quantum number s and g is:

$$\mu_z = g\mu_B s$$

3.2.6 Energy of Magnetic Moment

Consider a compass needle at rest, pointing north. If work is done to turn the needle around 180° so that it is held pointing south, then the needle has acquired **potential energy**. In a similar way, a magnetic moment which for some reason is not aligned with an external field within which it is located has potential energy – and so if it is released, it will acquire kinetic energy of rotation as it turns and aligns with the field. The potential energy depends upon the angle of the magnetic moment and the field and is over and above the potential energy of the electrons in their orbits or energy states. The **magnetic potential energy** is given by:

$$U = -\mu_M B \cos\theta \quad \text{or} \quad U = -\mathbf{\mu_M} \bullet \mathbf{B}$$

The potential energy is a maximum at $+\mu_M B$ when $\theta = \pi$ (aligned 180° against the field) and a minimum $-\mu_M B$ at $\theta = 0°$ (fully aligned with the field). The zero position of potential energy corresponds $\theta = \pi/2$ (90°) for convenience.

In general, for an electron the magnetic moment is:

$$\mu_M = g\left(\frac{-q_e}{2m_e}\right) L = g\mu_B l \text{ since } L = l\hbar; \quad \mu_B = \frac{-q_e\hbar}{2m_e}$$

The factor $g\frac{-q_e}{2m_e}$ is called the **gyromagnetic ratio**.

The maximum magnetic potential energy of μ_M occurs when L is aligned with the field (i.e., μ_M aligned against the field $\theta = \pi$):

$$U = g\left(+\frac{q_e}{2m_e}\right) LB$$

But the component of the **orbital angular momentum** L in the z direction can only take on discrete values given by the quantum number m. Thus, the magnetic potential energy of the *orbital* magnetic moment of an electron is restricted to certain maximum values given by:

$$U = \left(+\frac{q_e\hbar}{2m_e}\right) Bm \qquad m = -l, -l+1, ...l-1, l \qquad l = 0,1,2,...n-1$$

→ Bohr magneton

For **spin angular momentum** S, we have much the same situation where the magnetic potential energy of the *spin* magnetic moment of an electron is restricted to certain maximum values given by:

$$U = 2\left(+\frac{q_e\hbar}{2m_e}\right) Bs \qquad s = \frac{1}{2}, -\frac{1}{2}$$

3.2.7 Zeeman Splitting

Considering the allowable values of magnetic potential energy associated with magnetic moments of atoms in a magnetic field, it is of great interest to observe what effect the magnetic field has on the energies of electrons in atoms. That is, the energy associated with the magnetic moment (which exists when the electron finds itself in a magnetic field) either adds to or subtracts from the energy which it already has at that particular energy level in the atom, depending on the values of the quantum numbers l and s.

For the case of **orbital motion**, we can let, for example, $l = 1$ and m take on the values -1, 0 and 1. The allowable magnetic potential energies for an applied external field B are thus:

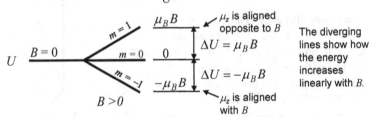

In the case of **spin**, the magnetic potential energy of the electron in the presence of a magnetic field depends upon the spin quantum number s.

U — $B = 0$ — $s = 1/2$ $\mu_B B$ — μ_z is aligned opposite to B (S_z aligned with B: spin up)

$s = -1/2$ $-\mu_B B$

$\Delta U = 2\mu_B B$ $\quad (g = 2)$

μ_z is aligned with B (S_z opposite to B: spin down)

$B > 0$

The splitting of energy into $2l + 1$ levels in the presence of a magnetic field is called **Zeeman splitting**.

Of course an electron in an atom contains both spin and orbital motion and so, in the presence of a magnetic field, there may be Zeeman splitting from both causes. Interaction between the two modes of splitting means that in practice, both orbital and spin motions should be considered together (i.e., $1 < g < 2$). More generally, we write:

$$\mu_z = g\mu_B \frac{J_z}{\hbar}$$

the z component of the total angular momentum $J_z = j\hbar$

$$= g\mu_B j$$

the total angular momentum quantum number

$$\boxed{U = g\mu_B B j}$$

For $g = 2$ and $j = 1/2$ $\mu_z = \mu_B$

3.2.8 Larmor Precession

When an atom is placed in a magnetic field, the electron energy levels change or split according to the value of the quantum numbers and the magnitude of the field. The change is always by an amount:

$$\Delta U = \pm \mu_B B \quad \text{where} \quad \mu_B = \frac{q_e \hbar}{2m}$$

If we consider the orbital motion of an electron in a magnetic field, then if the magnetic potential energy difference between one energy level and another is ΔU, this must also be equal to $h\nu$. That is, in a magnetic field, should there be a transition between the newly formed energy levels spaced ΔU apart, then the frequency of the associated photon is $E = h\nu$.

$$U = \left(+\frac{q_e \hbar}{2m_e} \right) Bm$$

$$E = \hbar\omega$$

$$\hbar\omega = \left(+\frac{q_e \hbar}{2m_e} \right) Bm$$

$$\boxed{\omega = \frac{q_e}{2m_e} Bm}$$

For the **spin** component where s is the spin quantum number,

$$U = 2\mu_B B s$$

$$\omega = \frac{q_e B s}{m_e}$$

Another physical interpretation of this frequency is that it represents a precession of the magnetic moment about B. This is called the **Larmor frequency**. That is, the magnetic moment rotates around B at the Larmor frequency. The frequency is proportional to the applied field strength; the stronger the field, the faster the magnetic moment rotates or precesses around B. This phenomenon forms the basis of procedures for measuring the value of g for a particular atomic system. The angle of precession is determined by the m quantum number.

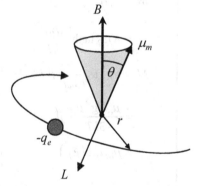

The Larmor precession arises because the magnetic moment is proportional to the angular momentum.

$$\mu_M = g\left(\frac{-q_e}{2m_e} \right) L$$

The precession is similar (except for quantum effects) to that of a spinning top or gyroscope.

3.3 Magnetic Properties

Summary

$$B = \mu_o H + \mu_o M$$ Net field in a material

$$\mu = \mu_o(1 + \chi)$$ Magnetic susceptibility

$$\mu_r = \frac{\mu}{\mu_o} = (1 + \chi)$$ Relative permeability

$$\Delta\mu_M = \frac{-q_e}{2m_e}\Delta L$$ Diamagnetism

$$M = \frac{N\mu_M^2 B}{3kT}$$ Classical paramagnetism

$$M = N\mu_B \frac{e^x - e^{-x}}{e^x + e^{-x}}$$ Quantum paramagnetism

$$= N\mu_B \tanh\left(\frac{\mu_B B}{kT}\right)$$

$$M = Ng^2\mu_B^2 \frac{B}{3kT} j(j+1)$$ Paramagnetism

$$\chi = \frac{\mu_o Ng^2\mu_B^2}{3kT} j(j+1)$$ Curie's law

$$\chi = \mu_o\mu_B^2 g(E_F)$$ Paramagnetic susceptibility (metals)

$$P_{rms} = \frac{1}{2}\omega H_{max}^2 \mu_o\mu_r \tan\delta$$ Power loss in a cored inductor

3.3.1 Permeability

When matter is placed in the region around a magnet or a current-carrying conductor, the magnetic field B in the space around the magnet or conductor is different from that which exists when the conductor is in a vacuum due to the **magnetisation** of the material.

1. Uniform field

2. Presence of material may concentrate field

This behaviour arises due to the interaction of the externally applied field and the internal field generated by the orbiting (and also spinning) electrons within the material.

Now, a magnetic field can be described as either B or H:

- The **magnetic induction** B ⟶ teslas (Wb m^{-2})
- The **magnetic field intensity** H

where: $\boxed{B = \mu H}$ amperes metre^{-1}

permeability

H is a measure of magnetism applied and is independent of the medium while B is a measure of magnetism, which results in the medium to which H is applied.

B and H have different physical significance.

The magnetic field intensity H is a field vector, and describes the magnetic field generated by a moving charged particle.

} The "magnetism" that is applied to a medium by some source

The magnetic induction B is a field vector which determines the magnetic force acting on a moving charged particle.

} The "magnetism" that is in turn created within the medium and is available to act upon anything placed within it

The resulting magnetic induction B produced by an external source of magnetic field H (such as a moving charge or an electric current) depends upon the "ease" with which the surrounding material permits the creation of magnetic field lines. This is the permeability of the medium. For free space, the permeability is $\mu_o = 4\pi \times 10^{-7}$ Wb A^{-1} m^{-1}

In free space the source of B is the applied field intensity H. When the space is occupied by a magnetic material, the field B now present is higher due to the extra contribution of magnetisation from the material. The total permeability is μ. The ratio μ / μ_o is the **relative permeability** μ_r.

3.3.2 Magnetic Materials

In most materials, the direction of magnetic moments is random and so there is usually no **net magnetic field** present (of course this is not the case for permanent magnets – which we will consider later).

When a material is placed within a magnetic field, unbalanced or net permanent magnetic moments within the material tend to align with and reinforce the external field. This alignment is opposed by the diamagnetic effect arising from the Lorentz force on the electron. The overall *net* magnetic moment per unit volume *induced* by the external field is called the **magnetisation** M.

$$M = \frac{\mu_{total}}{V}$$

We may then ask, what is the value of the resulting net magnetic field B within a material after an external field is applied? When a magnetic field is applied (such as through the motion of charges in a nearby conductor), it is done in terms of a magnetic field intensity H. In free space, the resulting magnetic induction B is given by $B = \mu_o H$.

When a material is present, there is an additional field created by the *net* magnetisation of the material. The additional field produced is proportional to the total **net magnetic moment** per unit volume induced in the material. The total net field B that now exists in the material is thus given by that which would be present in a vacuum, plus (or minus) the field from the magnetisation:

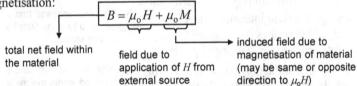

$$B = \mu_o H + \mu_o M$$

total net field within the material

field due to application of H from external source

induced field due to magnetisation of material (may be same or opposite direction to $\mu_o H$)

When M is in the same direction as H the material is said to be **paramagnetic** and the net field is greater than that originally applied as a result of the magnetisation of the material. When M is in the opposite direction from H, the net field B is weaker when the material is present and the material is said to be **diamagnetic**.

Why then do we need H if the contribution from the material can be included by the magnetisation M? H is used mainly because it is a macroscopic quantity and is more easily handled than the magnetisation M. In electrostatics, E and D are usually used rather than E and the polarisation P. Note, the force experienced by a charge in an electric field depends on E (not D) and the force experienced by a moving charge in a magnetic field depends on B (not H) and so E and B are usually of more interest in explaining physical phenomena.

3.3.3 B, H and M

The relationship between B, H and M provides a very useful picture of what happens when a material finds itself in a magnetic field. Consider the magnetic field B that exists within a material after it is placed in the field. What is the value of B at this condition?

This is the contribution to the resulting value of B arising from the ultimate source of the magnetic field (e.g., the current in a solenoid).
$$\mu_o H$$

H is the "cause" of the magnetic effects which follow.

$+$

This is the contribution to the resulting value of B arising from the magnetisation within the material.
$$\mu_o M$$

M results from alignment of dipoles within the material and the induced field $\mu_o M$ may oppose or reinforce the applied field depending on the nature of the material.

This is the resulting net field that is available to apply forces to any charges that happen to now move within it.
$$B$$

B is the combined "effect" resulting from the application of H and the magnetisation M.

Consider a wire-wound solid **toroid** with an air gap. Magnetisation of the material exists within the material of the toroid due to $\mu_o H$ from the current in the windings, and so we have both $\mu_o H$ and M vectors arriving at the surface of the gap. On the other side of the gap, M vectors are also to be found due to magnetisation of the toroid material, so what happens *in the gap*?

The M vectors end on the surface of the face of the gap and continue on into the gap as H vectors.

$\mu_o M \quad \mu_o H \quad \mu_o M$

$B = \mu_o H + \mu_o M$

$\mu_o H \quad \mu_o H \quad \mu_o H$

$B = \mu_o H$

$\mu_o M \quad \mu_o H \quad \mu_o M$

Note, there cannot be any M in the gap because there is no material there to be magnetised. H in the gap is larger than H in the material. B in the gap is the same as B in the material (neglecting fringing of the field in the gap).

3.3.4 Magnetic Susceptibility

The net **magnetisation** that occurs within a material placed in a magnetic field of intensity H depends upon the strength of H. That is, the magnetisation is proportional to H.

$$M \propto B_o$$

$$\boxed{M = \chi H}$$

magnetic susceptibility

thus: $B = \mu_o H + \mu_o M$

$\qquad = \mu_o (H + \chi H)$

$\qquad = \mu_o (1 + \chi) H$

$\qquad = \mu H$

$$\mu = \mu_o (1 + \chi)$$

$\mu_r = \dfrac{\mu}{\mu_o} \rightarrow$ relative permeability

$\qquad = (1 + \chi)$

The result is that the magnetic field within the space where the material is present is greater (or less depending on the sign of χ) by a factor $\mu_r = 1 + \chi$ than if there were a vacuum present.

Paramagnetic materials

Material	χ
oxygen	1.9×10^{-6}
aluminium	2.2×10^{-5}
platinum	2.6×10^{-4}

Diamagnetic materials

Material	χ
copper	-1.9×10^{-5}
gold	-3.6×10^{-5}
water	-9.0×10^{-5}

Note that the magnetic susceptibility is a small number (for both paramagnetic and diamagnetic materials) and so local field corrections (such as the **Clausius–Mosotti equation** used for electrical permittivity) can be ignored.

There is an important third class of materials where magnetic moments are aligned in **magnetic domains**, even in the absence of an external field. These are ferromagnetic materials (and have a large value of χ).

In summary:

- **diamagnetism** $\longrightarrow$ Magnetic moments add up in such a way
- **paramagnetism** as to oppose the external field.
- **ferromagnetism** Present in all atoms, but the effect is usually only observable in atoms whose shells are completely filled, since for unfilled shells, paramagnetism dominates.

Magnetic moments line up and serve to reinforce the external field.
Occurs in atoms whose shells are not completely filled.

Magnetic moments are strongly interacting with each other, lining up into **magnetic domains**, even when there is no external field present.

3.3.5 Diamagnetism

Consider an electron in a
circular orbit around a nucleus
with an angular velocity ω_0.

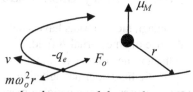

The **centripetal force** F_o
is due to the **Coulomb attraction** between the electron and the nucleus and is
directed inwards. This is balanced by a **centrifugal force** $m\omega_0{}^2 r$ acting
outward. With no external magnetic field applied, the initial **magnetic
moment** associated with the moving electron is:

$$\mu_M = IA = q_e \frac{\omega_0}{2\pi} \pi r^2 = \frac{q_e}{2}\omega_0 r^2$$

When a magnetic field is applied, an additional (Lorentz) force acts on the
electron: $F_L = qv \times B = -q_e vB$.

For the direction of B shown
here, the direction of the
Lorentz force is outwards and
opposite to the centripetal force.

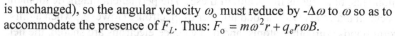

The centripetal force Fo does not
change (since the Coulomb attraction
is unchanged), so the angular velocity ω_0 must reduce by $-\Delta\omega$ to ω so as to
accommodate the presence of F_L. Thus: $F_o = m\omega^2 r + q_e r\omega B$.

Since the angular velocity is reduced by $-\Delta\omega$, the magnitude of the
magnetic moment is also reduced by:

$$\Delta\mu_M = \frac{-q_e}{2}\Delta\omega r^2$$

$$= \frac{-q_e}{2m_e}\Delta L$$

The significance of the
-ve sign is that $\Delta\mu_m$
points down in the
figure shown above.

This change in μ_M when divided by the
atomic volume, is the induced
magnetisation M of the atom brought
about by the presence of B. Note that the
degree of magnetisation depends upon
the strength of the applied field B (via
F_L). Since M is in the opposite direction
from B, the net field in the vicinity of
the atom is reduced. When the induced
magnetisation opposes the applied field,
the material is said to be **diamagnetic**.

If the motion of the electron were to be
counterclockwise, then μ_M would point
downwards in the above diagram, the
Lorentz force would point inwards, and
the angular velocity would increase.
$\Delta\mu_M$ would be positive (more
downwards), and so the induced
magnetisation would be again opposite
to that of the applied field. If the field B
had been applied downwards, and for
clockwise motion of the electron, then
the Lorenz force F_L would point inwards
towards the centre of rotation. In this
case, the angular velocity would
increase. The magnetic moment μ_M
would increase – and the change in μ_M
(now upwards) would again be
opposite to that of the applied field B
(now pointing downwards) .

3.3.6 Paramagnetism (classical view)

Paramagnetic materials have small permanent magnetic moments in the absence of an external B field.

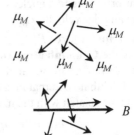

In the absence of an external magnetic field, thermal agitation of atoms causes the magnetic moments to be in a random direction.

When a field B is applied, the moments tend to preferentially line up in the direction of B. The net magnetic moment (the vector sum of all the atomic permanent moments) per unit volume is the **magnetisation** M induced by application of B. The amount of alignment, and hence the resulting magnetisation, depends on the competing effect of disorientation by thermal motion. From statistical mechanics the magnetisation is:

$$M = \frac{N\mu_M{}^2 B}{3kT}$$

Note that M is proportional to B and inversely proportional to T. In contrast, the magnetisation due to diamagnetism is proportional to B but independent of T.

We said above that paramagnetic materials have small permanent magnetic moments, but what is the origin of these? The permanent magnetic moments occur in materials with an odd number of electrons. However, when atoms form bonds with other atoms, and odd valence electrons are shared, the net magnetic moments arising from unpaired electrons cancel out. In a gas, such pairing may not occur and gases may exhibit paramagnetism. In solids, unfilled *inner* electron shells give rise to paramagnetism even though the outer valence electrons may be paired up in bonding.

In a classical sense, the permanent net magnetic moments of atoms of paramagnetic materials align with an applied field. The component of μ_M which is parallel to the field increases in magnitude as the moment rotates into alignment. Since the component in alignment with B increases in magnitude, the susceptibility must be positive and so the overall net field in the vicinity increases. When the induced magnetisation reinforces the applied field, the material is said to be **paramagnetic**.

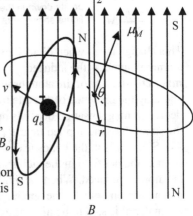

3.3.7 Paramagnetism (quantum view)

In general, the potential energy of the magnetic moment is expressed:

$$U = g\mu_B Bs$$

For an electron, with s taking on the values $-1/2$ and $+1/2$ and $g = 2$, in the presence of a magnetic field, **Zeeman splitting** gives us:

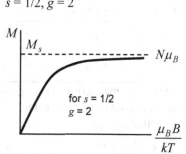

For a material exposed to a magnetic field, there will be N_1 atoms per unit volume in the lower energy level and N_2 atoms in the higher energy level, the total number of atoms per unit volume being $N = N_1 + N_2$. The two concentrations are related by the **Boltzmann distribution** such that:

$$\frac{N_2}{N_1} = e^{-\Delta U/kT}$$

From a macroscopic point of view, the **magnetisation** M gives the net total magnetic moment per unit volume of material, and so, in the presence of a field B aligned with the z axis, and for $s = 1/2$ and $g = 2$:

$$M = 2\mu_B(N_1 - N_2) \quad \text{(note, with } B = 0,\ N_1 = N_2 \text{ and } M = 0\text{)}$$

whereupon combining the above equations, we obtain an expression for the magnetisation from the average magnetic moment along the z axis:

$$M = N\mu_B \frac{e^x - e^{-x}}{e^x + e^{-x}} \quad \text{where } x = \frac{\mu_B B}{kT}$$
$$= N\mu_B \tanh\left(\frac{\mu_B B}{kT}\right)$$

$s = 1/2, g = 2$

M is proportional to B at low values of B (since when B is small, $\tanh x \approx x$) but at higher fields, eventually M reaches a **saturation** value of $M_s = N\mu_B$ whereupon all the moments are aligned with the field (spin down) and each contributes μ_B to the total magnetisation.

As the field is applied, the spacing between the Zeeman energy levels widens, and the concentration of moments in the lower energy level (aligned with the field) starts to increase, and so $N_1 > N_2$, thus leading to a non-zero value for M.

3.3.8 Total Angular Momentum

The total angular momentum of an atom is the vector sum of both spin and orbital motions: $\mathbf{J} = \mathbf{L} + \mathbf{S}$.

The possible values of the total angular momentum can be expressed in terms of a **total angular momentum quantum number** j and the magnitude of $\mathbf{J}$ is:

$$J = \sqrt{j(j+1)}\hbar$$

The z component of $\mathbf{J}$ is given by values of j from:

$$-j, -(j-1), -(j-2)...(j-2), (j-1), j$$

The largest value of the z component of $\mathbf{J}$ is, in accordance with the **uncertainty principle** $j\hbar$

The orbital and spin magnetic moments interact with each other and cause precession of the **L** and **S** angular momenta around **J**. The result is that the total net moment μ_M makes an angle with **J** due to the **Lande g factor** being greater for spin than for orbital motion.

Examples of values for z component of **J** in units of $\hbar$

j can either be an integer or a half-integer.

$j = 3/2$	$j = 2$	$j = 1/2$
+3/2	+2	−1/2
+1/2	+1	+1/2
−1/2	0	
−3/2	−1	
	−2	

electron
(a spin 1/2 particle)

Both the orbital and spin magnetic moments contribute to the magnetisation of materials. For non-paramagnetic materials, all these moments cancel because of pairing of electrons and so the net effect is **diamagnetism**. In paramagnetic materials, the magnetic moments from unpaired electrons do not ordinarily produce a net magnetic moment because of the disorientation of the moments due to thermal agitation. When an external magnetic field is applied, **Zeeman splitting** occurs and a net magnetisation develops along the direction of the field due to alignment of spin magnetic moments towards the lower energy state.

Now $\mu_M = g\mu_B \mathbf{J}$

Thus $$|\mu_M| = \mu_M = g\mu_B \sqrt{j(j+1)}$$

and so $$M = Ng^2 \mu_B^2 \frac{B}{3kT} j(j+1)$$

This is the magnetisation for general j and g *for small B*. The factor of 3 comes about due to averaging of the angular momentum over x, y and z directions in space.

For $j = 1/2$ and $g = 2$, we obtain: $M = N4\mu_B^2 \frac{B}{3kT}\frac{1}{2}\left(\frac{1}{2}+1\right)$

$$= N\mu_B^2 \frac{B}{kT} \text{ as before.}$$

3.3.9 Curie's Law

At temperatures above 0 K we can approximate M by:

$$M = N\mu_B \tanh\left(\frac{\mu_B B}{kT}\right) \quad j = 1/2, g = 2$$

$$\approx \frac{N\mu_B^2 B}{kT} \quad \text{since} \quad \frac{\mu_B B}{kT} \ll 1 \text{ at } T \gg 0 \text{ K.}$$

For a given applied H, the resulting magnetic field B is given by:

$$B = \mu_o(H + M)$$

$$\approx \mu_o H$$

since M is usually quite small. The ratio of M to H is the susceptibility:

$$\chi = \frac{M}{H}$$

The susceptibility is thus seen to be dependent on temperature:

$$\chi = \frac{N\mu_B^2 B}{kT} \frac{\mu_o}{B}$$

$$= \frac{N\mu_B^2 \mu_o}{kT}$$

More generally (for j and g), we have:

$$M = Ng^2\mu_B^2 \frac{B}{3kT} j(j+1)$$

$$\chi = \frac{M}{H}; B = \mu_o H$$

$$\boxed{\chi = \frac{\mu_o Ng^2\mu_B^2}{3kT} j(j+1)} \quad \textbf{Curie's law}$$

The **paramagnetic susceptibility** depends inversely on the temperature.

3.3.10 Paramagnetism in Metals

In general, metals are paramagnetic, although some exhibit
ferromagnetism. Paramagnetism arises from spin motion of the conduction
electrons, but it should be noted that the distribution of *conduction
electrons* follows the **Fermi–Dirac distribution**, not the **Boltzmann
distribution** used earlier.

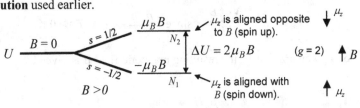

As before, **Zeeman splitting** results in electrons with spin up being raised in
energy and those with spin down being lowered in energy. In an attempt to
minimise the total energy, electrons *near the Fermi level* have the
opportunity to flip and transfer to the spin down, lower energy level. This
results in an imbalance of concentration of moments which are aligned with,
or opposite to the applied field and the result is of course a net
magnetisation of the material.

For each flip, the net
magnetisation increases by
$2\mu_B$. The magnetisation thus
depends upon the **density of
states** $g(E_F)$ at the Fermi
level. It is essentially
independent of temperature.

$$M = \frac{1}{2}g(E_F)\mu_B(2\mu_B)$$

$$= g(E_F)\mu_B^2 B$$

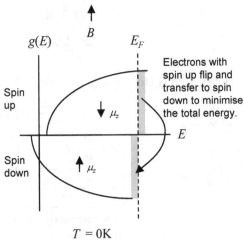

The **paramagnetic
susceptibility** is therefore:

$$\boxed{\chi = \mu_o\mu_B^2 g(E_F)}$$

The susceptibilities of metals are much
smaller than those of non-metals.

3.3.11 Complex Permeability

An ideal inductor consisting of an empty (air) coil has zero resistance and an impedance given by $Z_o = j\omega L$. When a magnetic core is inserted into the coil, the reactance increases:

$$Z_o$$

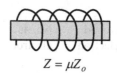

$$Z = \mu Z_o$$

In an AC circuit, hysteresis losses in the core lead to a resistance term R_L in the overall impedance of the coil (that is, the reactance term ωL in the impedance is out of phase with the loss term R_L). (The resistance term is within the core material and may be added, if desired, to the DC winding resistance of the coil.) Thus:

$$Z = R_L + jwL$$

$$R_L \qquad L_o$$

Now, let $\mu' = \dfrac{L}{L_o} = \mu_r$

and $\mu'' = \dfrac{R_L}{L_o \omega}$

$$Z = u'' L_o \omega + \mu' L_o j\omega$$
$$= j\omega L_o \left(\mu' - j\mu'' \right)$$

Complex permeability

Now, the loss factor is the ratio of the permeabilities:

$$\tan \delta = \frac{\mu''}{\mu'} = \frac{R_L}{\omega L}$$

The loss resistance is: $\quad R_L = \mu'' L_o \omega$

Power dissipation: $\quad P_{rms} = I_{rms}^2 R_L$
$$= I_{rms}^2 \mu'' L_o \omega$$
$$= \omega H_{rms}^2 \mu_o \mu''$$
$$= \omega H_{rms}^2 \mu_o \mu' \tan \delta$$

$$\boxed{P_{rms} = \frac{1}{2} \omega H_{max}^2 \mu_o \mu_r \tan \delta} \qquad \text{since } \mu' = \mu_r$$

3.3.12 Saturation Moment

The table below compares the electronic properties of some important magnetic materials assuming isolated atoms and no 4s moment contribution.

Element	No. 3d electrons	No. 4s electrons	No. unpaired 3d electrons	s	M_s (in μ_B)
Ti	2	2	2	1	2
Ti^{3+}	1	0	1	1/2	1
Ti^{4+}	0	0	0	0	0
Cu	10	1	0	0	0
Cu^{2+}	9	0	1	1/2	1
Fe	6	2	4	2	4
Fe^{2+}	6	0	4	2	4
Co	7	2	3	3/2	3
Co^{3+}	6	0	4	2	4
Ni	8	2	1	1	2
Ni^{2+}	8	0	1	1	2

$$M_s = (gS)\mu_B$$
$$g = 2 \quad \text{for spin}$$
$$s = \frac{\text{No. unpaired 3d electrons}}{2}$$

3.3.13 Example

If a single atom has a net magnetic moment of $\mu_M = 2.1 \times 10^{-23}$ A m^2, calculate the magnetisation and the magnetic induction for this material if all the moments are aligned with each other and the number density of atoms is 5×10^{27} atoms per m^3.

Answer:

If all the moments are aligned, then the magnetisation has reached the saturation level and so $M_s = N\mu_M$.

$$\mu_M = 2.1 \times 10^{-23} \, \text{Am}^2$$
$$N = 5 \times 10^{27} \, \text{m}^{-3}$$
$$M_S = \mu_M N$$
$$= 2.1 \times 10^{-23} \left(5 \times 10^{27} \right)$$
$$= 105 \times 10^3 \, \text{A m}^{-1}$$

With no external field, we have $H = 0$:

$$B = \mu_o H + \mu_o M$$
$$= \mu_o M$$
$$= 4\pi \times 10^{-7} \left(105 \times 10^3 \right)$$
$$= 0.132 \, \text{T}$$

3.3.14 Example

Consider the current loop created by the orbital motion of an electron. If the radius of the orbit is 0.5 Å, calculate the current for one Bohr magneton of magnetic moment. Also, calculate the angular velocity of the electron.

Answer:

The current is coulombs per second and thus the frequency can be found by dividing the current by the electronic charge.

$$r = 0.5 \times 10^{-10} \, m$$

$$\mu_B = IA$$

$$0.927 \times 10^{-23} = I\left(\pi\right)\!\left(0.5 \times 10^{-10}\right)^2$$

$$I = 1.18 \times 10^{-3} \, A$$

$$I = -q_e v$$

$$1.18 \times 10^{-3} = -1.6 \times 10^{-19} v$$

$$v = 7.38 \times 10^{15} \, Hz$$

3.3.15 Example

Calculate the energy required to rotate an atomic magnetic moment with a spin $j = 2$ from being aligned with an external field B to be aligned completely against the field. Then calculate the ratio of moments which are aligned with the field to those against at 300 K and 150 K.

Answer:

For $j = 2$, there are $(2j+1) = 5$ levels, and so four ΔE's.

$$j = 2$$
$$B = 0.4 \text{ T}$$
$$\Delta E = g\mu_B B$$
$$= 4(g\mu_B B)$$
$$= 4(2)(9.27 \times 10^{-24})(0.4)$$
$$= 2.966 \times 10^{-23} \text{ J}$$
$$= 1.85 \times 10^{-4} \text{ eV}$$

$$\frac{N_2}{N_1} = e^{-\Delta E / kT}$$

$$= e^{-\frac{1.85 \times 10^{-4}}{0.026}}$$

$$= 0.993 @ 300K$$ At 300K, we have almost equal numbers of moments aligned with and against the field.

$$\frac{N_2}{N_1} = e^{-\frac{1.85 \times 10^{-4}}{0.013}}$$

$$= 0.986 @ 150K$$ As the temperature is lowered, N_2 (where moments are aligned against the field) becomes less.

3.4 Ferromagnetism

Summary

$$\lambda = \frac{H_e}{M} = \frac{kT_C}{\mu_o N \mu_B^2}$$ Weiss constant

$$\mu_M = g\mu_B \sqrt{j(j+1)}$$ Saturation flux

$$M_s = N\mu_M$$

$$\chi = \frac{C}{T - T_C}$$ Curie-Weiss law

$$C = \frac{N\mu_B^2 \mu_o}{k}$$

$$P_{rms} = \frac{1}{2}\mu_r \mu_o H_{max}^2 \omega \tan\delta$$ Hysteresis power loss

$$\mu_r = \mu_{ro}\frac{1}{\left(1 + \omega^2 \tau^2\right)}$$ High frequency response of ferrites

$$\tan\delta = \omega\tau$$

3.4.1 Ferromagnetism

Ferromagnetic materials have permeabilities much larger than that of free space. μ_r = 1000 to 10000.

The spin magnetic moments within a ferromagnetic material interact with each other even when there is no external field present by a quantum mechanical **exchange coupling force**. These interactions cause the creation of a strong internal **molecular magnetic field**. This field causes neighbouring moments to align themselves parallel to each other in regions called **magnetic domains** of size approximately 10^{-7}m.

With no external field, the orientations of the *domains* are random. Within a domain, the orientation of the magnetic moments is aligned and the magnetisation is at the saturation level. When an external field H is applied, the domains tend to reorient so that the effect of this is that those domains already in alignment with the field tend to grow in size at the expense of others, not aligned, which shrink. As H is increased, a **saturation** point is reached where all domains contain magnetic moments that are aligned with the field.

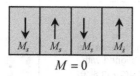 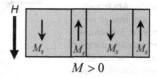

Iron is the most significant magnetic material hence the term **ferromagnetic**, but the term also refers to other elements such as nickel and cobalt.

In the presence of an external field H, there thus exists a *net* magnetisation M – the material becomes magnetised. If the aligned domains remain aligned upon removal of the external field, we have created a **permanent magnet**. It should be noted, however, that although the magnetic field of a permanent magnet can be used to perform mechanical work, we have to put work into the system first to align the domains.

3.4.2 *B-H* Curve

With a paramagnetic material, the application
of H results in a field B in a linear fashion
such that: $B = \mu_o H + \mu_o M$

$\qquad\qquad = \mu H$

When H is applied to a ferromagnetic material, the relationship is not
linear. The **permeability** (the ratio of B/H at any point – and not the
instantaneous slope) is a function of H.

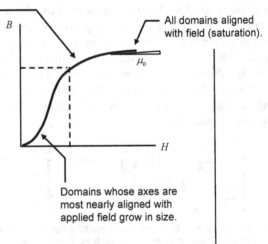

The permeability of a
ferromagnetic material
varies as the external
magnetic field intensity H
is varied. For a given
value of H, B depends on
μ because H is great
enough to force *domains*
to align themselves with
H rather than their
preferred crystalline
orientation. Alternately
we can say that μ
depends on B because
the permeability depends
upon what fraction of the
magnetic domains has
aligned its moments with
the magnetic field.

All domains aligned
with field (saturation).

Domains whose axes are
most nearly aligned with
applied field grow in size.

In a ferromagnetic material, an applied H aligns the *domains* –
not so much the magnetic moments. Within a domain, the
magnetic moments are already aligned by the strong internal
molecular field. All H does in a ferromagnetic material is to
align the *domains* in a particular direction.

At **saturation**, all the domains, and hence all the magnetic
moments, are aligned. The net magnetisation within the material
reaches a saturation value M_s. The relationship between B and H at
saturation is similar to that of a paramagnetic material but in this
case, the magnetisation M_s remains at its saturation value:

$$B = \mu_o H + \mu_o M_s$$

Above saturation, any increase in B due to an increase in H is due
to $\mu_o H$.

3.4.3 Hysteresis

After a ferromagnetic material has been magnetised by a field intensity H, when H is brought to zero, most ferromagnetic materials will show some residual net magnetic field B. A **demagnetising field** H_c is required to be applied to reduce B to zero.

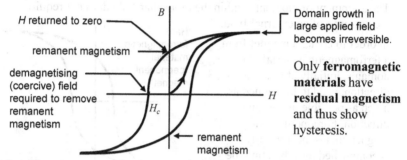

H returned to zero

remanent magnetism

demagnetising (coercive) field required to remove remanent magnetism

H_c

remanent magnetism

Domain growth in large applied field becomes irreversible.

Only **ferromagnetic materials** have **residual magnetism** and thus show hysteresis.

When the magnetic field intensity H arises from an alternating current, the flux density B in a ferromagnetic material tends to lag behind the magnetic field intensity H which creates it. This is called **hysteresis loss**.

The area within the **hysteresis loop** represents an energy loss as the field has to be reversed to negate the residual magnetism arising from the irreversible alignment of domains at high values of H.

Alternate reversals of the external field, such as in an AC circuit, can therefore result in significant hysteresis losses. The energy dissipated per unit volume is the integral:

$$U_H = \oint H dB$$

The surface integral symbol is used to denote an integration over one complete cycle.

For a frequency ω, the rms power dissipation within a volume V of material is:

$$P_{rms} = \frac{\omega}{2\pi} V \oint H dB$$

The loss term manifests itself as a complex permeability, where the power loss is expressed in terms of the loss factor $\tan \delta$:

$$\boxed{P_{rms} = \frac{1}{2} \mu_r \mu_0 H_{max}^2 \omega \tan \delta}$$

3.4.4 Ferromagnetic Materials

Examples of ferromagnetic materials are Fe, Co and Ni. Materials made from rare earth elements are in some cases also ferromagnetic. Ferromagnetism can occur in both metals and insulators. The origin of ferromagnetism is the presence of a strong internal **molecular magnetic field** that acts so as to align the spin moments of electrons into domains. The alignment of moments within the molecular field does not require the presence of an external field.

In order to create a magnet from a ferromagnetic material, the domains, initially at random direction, need to be aligned as much as possible in the same direction. For a permanent magnet, this is most usually accomplished by subjecting the material to a strong external magnetic field. For permanent magnets, the domains, once aligned, are difficult to reorient. These materials are called **hard ferromagnetic materials**. However, it should be noted that although a strong internal molecular field is required to orient magnetic moments inside a domain, alignment of domains into a single direction need not require a strong external field. In some materials, even a small field will do. In **soft ferromagnetic materials**, the direction of orientation of domains may be very easily accomplished. When the external field is removed from a soft ferromagnetic material, very little residual net magnetisation exists.

Hard ferromagnetic material (permanent magnet)

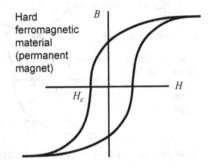

Soft ferromagnetic material (transformer core)

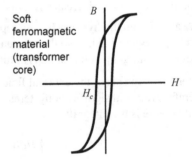

Linear 2nd quadrant materials (NdFeB)

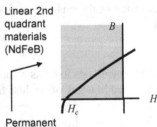

Permanent magnets are usually operated in the 2nd quadrant.

3.4.5 Molecular Field

Although we have mentioned the strong **molecular field** that exists in ferromagnetic materials several times, we have not yet made any real explanation for why it exists. For now, we can say that the molecular field H_e is created by spontaneous alignment of magnetic moments μ_M in the material, even in the absence of an externally applied field. It is reasonable to assume that the strength of the molecular field H_e is proportional to the spontaneous magnetisation M:

— **Weiss constant**

$$H_e = \lambda M$$

It will be later shown that the **molecular exchange field** B_e is very much larger than any externally applied H and is formed predominantly by the magnetisation of the material: $B_e = \mu_o(H + \lambda M)$

$$\approx \mu_o \lambda M$$

For $g = 2$ and $j = 1/2$ (for convenience) we have:

$$M = N\mu_B \tanh\left(\frac{\mu_B B_e}{kT}\right) \quad \text{curved line}$$

$$B_e = \mu_o \lambda M$$

$$M = N\mu_B \tanh\left(\frac{\mu_B \mu_o \lambda M}{kT}\right)$$

If we let $x = \dfrac{\mu_B \mu_o \lambda M}{kT}$ then

$$M = \frac{kT}{\mu_B \mu_o \lambda} x \quad \text{straight line}$$

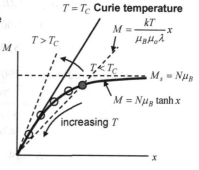

The point of intersection of the straight and curved lines gives consistent values of M for different values of T for a condition of **spontaneous magnetisation.** If T starts off low, then the point of intersection is over to the right hand side of the figure. As T is raised, the magnetisation M becomes lower as the straight line gets steeper and the point of intersection slowly, and then more rapidly, approaches 0, at which time $T = T_C$. At $T > T_C$, there is no spontaneous magnetisation.

At $T = T_C$, and for small values of x, where $M = N\mu_B\left(\dfrac{\mu_B \mu_o H_e}{kT}\right)$

we obtain: $\dfrac{M}{H_e} = \dfrac{N\mu_o \mu_B^2}{kT_C}$

and: $\boxed{\lambda = \dfrac{H_e}{M} = \dfrac{kT_C}{\mu_o N\mu_B^2}}\ g = 2, j = 1/2$

Weiss constant

3.4.6 Saturation Flux Density

The **saturation magnetisation** for a ferromagnetic material can be expressed in terms of the magnetic moment per atom:

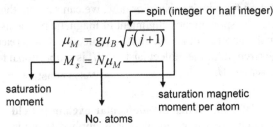

spin (integer or half integer)

$$\mu_M = g\mu_B\sqrt{j(j+1)}$$
$$M_s = N\mu_M$$

saturation moment

No. atoms

saturation magnetic moment per atom

Given the saturation moment, we are able to calculate the **saturation flux density** B_s as follows. For example, for iron:

$$\text{if } \mu_M = 2.2\mu_B$$
$$\rho = 7900\,\text{kgm}^{-3}$$
$$m.w. = 55\times10^{-3}\,\text{kg/mol}$$

Then in $1m^3$ iron, we have: $= 7900\,\text{kg}$

$$n = \frac{7900}{55\times10^{-3}}$$
$$= 143636\,\text{mol}$$
$$N = 143636\left(6.023\times10^{23}\right)$$
$$= 8.65\times10^{28}\,\text{atoms}$$
$$M_s = 8.65\times10^{28}(2.2)\left(9.27\times10^{-24}\right) \quad \text{saturation}$$
$$= 1.76\times10^{6}\,\text{A m}^{-1} \quad \text{magnetisation}$$
$$B_s = \mu_o M_s \quad \text{saturation flux}$$
$$= 4\pi\times10^{-7}\left(1.76\times10^{6}\right) \quad \text{density}$$
$$= 2.22\,\text{T}$$

3.4.7 B of the Molecular Field

The flux density for the exchange force **molecular field** B_e can be calculated if the magnetisation and T_C are known. For example, for iron, if $T_c = 1043$K and the saturation magnetisation per atom $M_s = 2.2\mu_B$, then:

$$H_e = \lambda M_s$$

$$M_s = N\mu_M$$

$$= Ng\mu_B j \quad \text{where } N = 1 \text{ for one atom}$$

Then:
$$\lambda = \frac{3kT_C}{\mu_o Ng^2\mu_B^2 j(j+1)}$$

$$B_e = \mu_o \lambda M$$

$$= \frac{\mu_o 3kT_C M}{\mu_o Ng^2\mu_B^2 j(j+1)}$$

$$= \frac{\mu_o 3kT_C Ng\mu_B j}{\mu_o Ng^2\mu_B^2 j(j+1)}$$

$$= \frac{3kT_C}{g\mu_B(j+1)}$$

$$= \frac{3kT_C}{g\mu_B j + g\mu_B}$$

$$= \frac{3(1.38\times10^{-23})1043}{(2.2+2)(9.27\times10^{-24})} \quad \text{since } T_C = 1043 \text{ and } M_s = 2.2\mu_B$$

$$= 1109 \text{ T}$$

Compare this with the saturation flux density B_s on the previous page. B_e is a very strong field.

3.4.8 Curie–Weiss Law

Above the **critical temperature**, the molecular field disappears and the material is paramagnetic. As T is raised and approaches T_C, the normally very strong molecular field becomes comparable to that of H which may be applied externally, and so the field B becomes:

$$B = \mu_o(H + \lambda M)$$

$$M \approx N\mu_B\left(\frac{\mu_B B_{local}}{kT}\right) \qquad g = 2, j = 1/2$$

$$= N\mu_B\left(\frac{\mu_o\mu_B(H + \lambda M)}{kT}\right)$$

$$\frac{M}{M_s} \approx \frac{\mu_o\mu_B(H + \lambda M)}{kT} \qquad \text{where } M_s = N\mu_B$$

$$= \frac{\mu_o\mu_B H}{kT} + \frac{\mu_o\mu_B M}{kT}\frac{kT_C}{\mu_o N\mu_B^2} \qquad \text{since } \lambda = \frac{kT_C}{\mu_o N\mu_B^2}$$

$$= \frac{\mu_o\mu_B H}{kT} + \frac{M}{M_s}\frac{T_C}{T}$$

$$\frac{M}{M_s}\left(1 - \frac{T_C}{T}\right) = \frac{\mu_o\mu_B H}{kT}$$

$$\frac{M}{M_s} = \frac{\mu_o\mu_B H}{kT}\left(\frac{T}{T - T_C}\right)$$

$$= \frac{\mu_o\mu_B H}{k(T - T_C)}$$

$$\chi = \frac{M}{H}$$

$$= \frac{\mu_o\mu_B M_s}{k(T - T_C)}$$

$$\boxed{\chi = \frac{C}{T - T_C}}$$

where $C = \dfrac{N\mu_B^2\mu_o}{k}$

In a ferromagnetic material initially at a temperature $T > T_C$, the material is paramagnetic. As T is lowered and approaches T_C, χ increases and diverges at $T = T_C$. This marks the transition to the ferromagnetic state.

Compare with $\chi = \dfrac{N\mu_B^2\mu_o}{kT}$

3.4.9 Exchange Energy

The pairing of outer electrons in neighbouring atoms into opposite directions means that for most materials, there is no net magnetic moment. When an external field is applied, **Zeeman splitting** results in most solid materials exhibiting **paramagnetism**.

The preferential alignment of electron spins into anti-parallel pairs changes with the distance between the electrons. When two electrons are far away from each other, they can have any orientation with respect to each other. As they become closer together, it is energetically favourable for their spins to align with each other. When they are close together, it is energetically preferable for their spins to be opposite to each other.

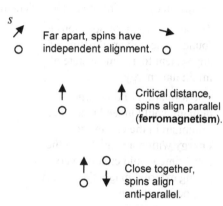

Far apart, spins have independent alignment.

Critical distance, spins align parallel (**ferromagnetism**).

Close together, spins align anti-parallel.

For ferromagnetic materials, the outer electron shells are filled and form anti-parallel pairs within each atoms. However, for these materials, there are unpaired electrons in inner shells, and these can interact with their counterparts in neighbouring atoms. If the distances are just right, then there is an alignment of the spins and magnetic moments of these inner electrons between atoms, thus leading to the formation of domains whose magnetic moments are aligned.

The nature of the alignment of spins arises due to the character of a quantum-mechanical **exchange force**. This force, initially very low when the electrons are far apart, increases as the two electrons are brought closer together and reaches a maximum at a certain critical distance between two electrons – in which the spins lead to a lower exchange energy when they are in the same direction. The exchange force decreases again as the electrons are brought even more closely together and becomes negative at which point it is more favourable for the electrons to have their spins in opposite directions. In ferromagnetic materials, the unpaired inner electrons find themselves at this critical distance with other unpaired inner electrons in neighbouring atoms and so the magnetic moments tend to become aligned.

3.4.10 Domain Formation

If the spacing between unpaired inner electrons in ferromagnetic materials leads to the preferential alignment of magnetic moments of atoms, we may then ask why do we not obtain one spontaneously formed domain. Why are there (in the absence of an external field) a profusion of domains at random orientation?

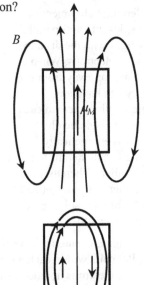

The answer to these questions is found in the universal nature of any system to assume a state of minimum energy.

The formation of one large domain would indeed result in a minimum in the **exchange energy** within a solid, but at the same time would create a very strong external magnetic field. A magnetic field has **magnetostatic energy**.

If the material were to divide into, say, two domains, this would reduce the external magnetic field and lower the magnetostatic energy at the expense of only a slight increase in the total exchange energy.

A minimum in the magnetostatic energy arises when there is a large number of domains with random orientation. Thus, we have competing conditions for a minimum total energy. The formation of random domains would result in a lower magnetostatic energy while the formation of one large domain would result in a minimum in the exchange energy.

There are other energies which also come into play – such as **magnetostriction** and **anisotropy** energies. The resulting sum of all these competing effects results in an energy minimum with the formation of a large number of randomly oriented domains.

3.4.11 Bloch Wall

The **exchange force** is a short range force. For ferromagnetic
materials, where unpaired inner shell electrons are affected by the
exchange force from similar electrons in neighbouring atoms, the
exchange energy is the greatest when the two electrons have opposite
spin. That is, the **exchange energy** is the greatest at the boundary
between two domains.

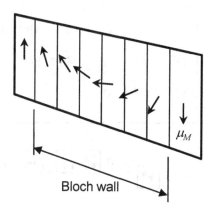

We must have the
formation of domains to
minimise the external
magnetostatic energy, but
at the same time, it is
preferable to also minimise
the exchange energy at the
domain walls. Nature does
this in ferromagnetic
materials by creating a
gradual transition of spin
over a short distance, the
Bloch wall, so that the
exchange energy is
minimised over to adjacent
atoms while at the same
time achieves the creation
of a new domain.

Bloch wall

Because of this gradual transition in magnetic moment from one domain to
the other, the exchange energy associated with each neighbouring pair of
atoms is nearly at the lowest possible level – and so the region of transition
can be easily moved under the application of an external field H. Domains
grow under the application of an applied field intensity H by movement of
the Bloch wall. This is why ferromagnetic materials can be easily
magnetised – it doesn't take much applied H to move a Bloch wall along
within the material and grow a domain. In hard magnetic materials, wall
movement is made harder due to imperfections, anisotropy and inclusions
in the material and, once moved, are difficult to reverse.

3.4.12 Unpaired 3d Electrons

The number of unpaired 3d electrons is a very important quantity in determining the magnetic properties of materials. The diagram below gives a useful summary for various ions of interest.

No unpaired 3d electrons

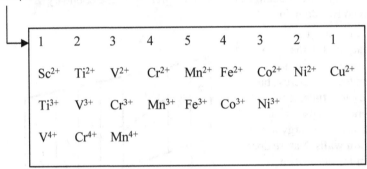

1	2	3	4	5	4	3	2	1
Sc^{2+}	Ti^{2+}	V^{2+}	Cr^{2+}	Mn^{2+}	Fe^{2+}	Co^{2+}	Ni^{2+}	Cu^{2+}
Ti^{3+}	V^{3+}	Cr^{3+}	Mn^{3+}	Fe^{3+}	Co^{3+}	Ni^{3+}		
V^{4+}	Cr^{4+}	Mn^{4+}						

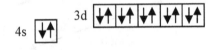

The 4s energy level can hold up to 2 electrons, with opposite spins. The 3d energy level can hold up to 10 electrons.

Let's consider Fe. Fe has $2 \times 4s$ electrons and $6 \times 3d$ electrons.

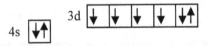

Electrons fill levels according to **Hund's rules**. Thus, there are 4 unpaired 3d electrons for Fe.

For Fe^{2+}, we have still 4 unpaired 3d electrons as the 4s electrons are lost to form the 2+ ion.

For Fe^{3+}, we lose one of the 3d electrons and so we obtain 5 unpaired 3d electrons.

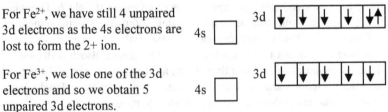

The saturation magnetic moment (due to spin) is equal to the number of unpaired 3d electrons $\times \mu_B$.

3.4.13 Ferromagnetism in Metals

So far, our description of ferromagnetism has relied on 3d electrons with unpaired spins being located at fixed positions in relation to each other (i.e., at specific sites in a crystalline lattice). This is fine for **insulators**, but in **metals**, outer shell conduction electrons readily move throughout the crystal structure and so the condition for the critical distance for alignment of the magnetic moments may not always be satisfied by one particular electron pair.

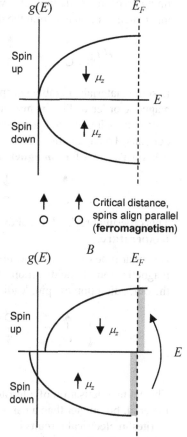

For metals, we recognise that the electrons of interest as far as **ferromagnetism** goes occupy the 3d band. Before exchange forces come into play, the density of states of the two orientations of spin can be considered equal.

When the **exchange force** occurs, magnetic moments tend to align in the up direction. Electrons transfer from the spin down to the spin up state in an attempt to lower the total energy of the system (note: this is opposite to the situation in paramagnetism). Doing so means that the energies of the states for the spin up and spin down states are not equal:

For each flip, the net magnetisation increases by $2\mu_B$. The amount of magnetisation depends upon the strength of the exchange force. The electrons which transfer from spin down to spin up gain kinetic energy which equals the loss in exchange energy. The condition for ferromagnetism is:

$$2\mu_o \lambda \mu_B^2 > \frac{2}{g(E_F)}$$

This condition is most easily met when the **Weiss constant** λ is large and the **density of states** at E_F is large. A large density of states indicates a narrow band within which can be accommodated a relatively large number of electrons – precisely the properties of the 3d band in metals such as Fe, Co and Ni.

3.4.14 Ferrimagnetism

In **paramagnetic materials**, in the absence of an externally applied field, the magnetic moments within the material are in a random direction and there is no net magnetisation.

In **ferromagnetic materials**, the molecular field causes spontaneous net magnetisation over small volumes or domains in the material. Thus, we can say that within a domain, there is some order to the magnetic effects – the lining up of magnetic moments where the spins of electrons in unfilled inner shells of atoms within the domain all point in the same direction.

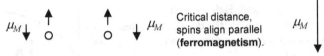

In some materials, notably **compounds** involving transition metals, there is magnetic order within the material where the spins of adjacent atoms, instead of being aligned, are opposite. These atoms are not aligned into domains. The magnetic ordering takes place usually within a unit cell. This is known as **anti-ferromagnetism**. The overall net magnetisation is zero.

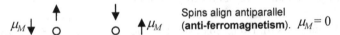

This ordering breaks down above a critical temperature called the **Neel temperature**.

A similar type of ordering occurs in **ferrimagnetic materials** but the magnetisation of one direction of spin is less than the magnetisation of the other direction of spin, leading to a net magnetisation.

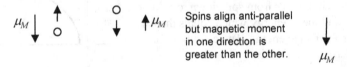

The net magnetisation of a ferrite is much lower than a ferromagnetic material, but higher than a paramagnetic material. The main advantage of ferrites in electrical circuits is that they are insulators, and so may operate with periodically reversing magnetic fields in **high frequency** applications without the eddy current losses that would be experienced with conducting ferromagnets.

3.4.15 Ferrites

Ferrites are iron oxides with a chemical formula: $X^{2+}O^{2-}Fe^{3+}{}_2O^{2-}{}_3$ where X is one of Zn^{2+}, Cu^{2+}, Ni^{2+}, Co^{2+}, Mn^{2+}, Fe^{2+}.

The ions are arranged in an interlocking combination of two crystal lattice geometries A and B. Ions at the A sites have **tetrahedral** oxygen arrangements and those at the B sites have **octahedral** structure. The structure of some ferrites can be summarised as follows:

Ferrite	A (tetrahedral)	B (octahedral)	$\mu_M = N\mu_B$
$Ni\,Fe_2O_4$	Fe^{3+}	$Fe^{3+}Ni^{2+}$	$2\,\mu_B$
$CoFe_2O_4$	$Fe^{3+}{}_{0.8}Co^{2+}{}_{0.2}$	$Fe^{3+}{}_{1.2}Co^{2+}{}_{0.8}$	$3.8\mu_B$
$Zn_{0.3}Ni_{0.7}Fe_2O_4$	$Zn_{0.3}Fe^{3+}{}_{0.7}$	$Ni^{2+}{}_{0.7}Fe^{3+}{}_{1.3}$	$4.4\mu_B$
$MgFe_2O_4$	$Fe^{3+}{}_{0.9}Mg^{2+}{}_{0.1}$	$Fe^{3+}{}_{1.1}Mg^{2+}{}_{0.9}$	$1\,\mu_B$

The magnetic moments of the magnetic ions on the A sites point in the opposite direction to those of the magnetic ions at the B sites. These are called **inverse ferrites**. The net magnetic moment for one molecular unit is found from a weighted sum.

Element	No. unpaired 3d electrons
Fe^{2+}	4
Fe^{3+}	5
Co^{2+}	3
Ni^{2+}	2

For example an Fe^{3+} ion has 5 unpaired 3d electrons. The magnetic moment of this ion is thus $5\mu_B$. The Ni^{2+} ion has 2 unpaired electrons, and so the magnetic moment is $2\mu_B$. For the ferrite $NiFe_2O_4$, the net magnetic moment, is $5\mu_B - (5+2)\,\mu_B = 2\mu_B$. For the ferrite $CoFe_2O_4$, the net magnetic moment would be $((0.8)5+(0.2)3 - (1.2)5+(0.8)3)\,\mu_B = 3.8\mu_B$.

Each unit cell has 8 molecular units. The **saturation magnetisation** M_s and flux density B_s for one unit cell can be calculated if the size of the unit cell is known (e.g., as given by x-ray analysis). For example, for $NiFe_2O_4$, if the size of the unit cell is measured as 8.34×10^{-10} m, then:

$$M_s\,/\,molecule = \frac{8(2)\mu_B}{\left(8.34 \times 10^{-10}\right)^3}$$

$$= 2.56 \times 10^5 \,A\,m^{-1}$$

$$B_s = \mu_o M_s$$

$$= 0.32\,T$$

An interesting example of a ferrite is the mineral **magnetite** ($Fe^{2+}O^{2-}\,Fe^{3+}{}_2\,O^{2-}{}_3$) with a moment of $4\mu_B$. The significance of this is that **lodestone** (magnetite) is a ferrite, and not a ferromagnet.

A **normal ferrite** (as distinct from an inverse ferrite) has no magnetic moment at the A sites. For example, $ZnFe_2O_4$ has non-magnetic Zn^{2+} ions on the A sites and the two Fe^{3+} ions are at the B sites whose moments point in opposite directions (an anti-ferromagnet).

3.4.16 Mixed Ferrites and Garnets

It is desirable to have the largest possible net magnetic moment for a ferrite and this can sometimes be achieved by a mixture of normal and inverse ferrite materials. The objective is to reduce the cancellation of the magnetic moments of the Fe^{3+} ions on the A and B sites. Consider the following mixed ferrite:

$$(ZnOFe_2O_3)_x(MnOFe_2O_3)_{1-x}$$

A site   B site

$$Zn^{2+}{}_xFe^{3+}{}_{1-x} \qquad Mn^{2+}{}_{1-x}Fe^{3+}{}_{1+x}$$

The net magnetic moment is:

$$N\mu_B = 5(1+x)+5(1-x)-5(1-x)$$

The maximum magnetic moment occurs when $x = 1/2$.

Note, we don't include Zn^{2+} because this ion has no moment.

$$N\mu_B = 5(1.5)+5(0.5)-5(0.5)$$
$$= 7.5+2.5-2.5$$
$$= 7.5$$

In practice, above $x = \frac{1}{2}$, anti-ferromagnetism on the B sites reduces the net magnetic moment over that predicted by the above formula. That is, at $x = 1$, we have $ZnFe_2O_4$ where non-magnetic Zn^{2+} ions on the A sites and the two Fe^{3+} ions (whose moments point in opposite directions) are at the B sites and so it is an anti-ferromagnet.

In the **mixed ferrite** shown above, the presence of the non-magnetic Zn^{2+} ion tends to thus reduce the cancelling effect of the opposite spins of the Fe^{3+} ions, thus leading to $\mu_M > 5\mu_B$.

Even greater moments can be obtained in **garnets**. Moments $> 9\mu B$ can be obtained (e.g., YIG).

3.4.17 High Frequency Response of Ferrites

Ferrites have considerable practical importance because of their high resistivity together with their magnetic properties. Ferrites are often used as the core in inductors where the wire is wound around the ferrite. Ferrites with low resistivity suffer from losses from the generation of **eddy currents**. Mn-Zn ferrites exhibit significant eddy current losses <100 kHz, while Ni-Zn ferrites have very little eddy current losses up to optical frequencies.

Experimentally, the response of a ferrite over a range of frequencies can be measured using a simple resonant Q meter circuit.

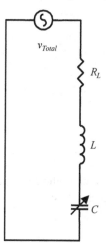

A copper conductor is wound around a cylindrical ferrite core. R and L are the resistance (loss) and inductance (complex) properties of the coil.

C is an adjustable capacitor of known capacitance.

As the frequency of the applied voltage to the coil is varied, a resonance condition will occur when at ω_R when $X_C = X_L$ and so:

$$X_C = \frac{1}{\omega C}$$

$$X_L = \omega L$$

$$\omega_R = \frac{1}{\sqrt{LC}}$$

The **Q factor** measures the sharpness of the **resonant peak** and is given by:

$$Q = \frac{\omega_R L}{R} = \frac{1}{R \omega_R C} = \frac{1}{R}\sqrt{\frac{L}{C}}$$

Instrumentation usually allows an adjustment for C and provides values of Q directly, from which the **loss factor** tan δ is determined from $1/Q$ and the inductance L from:

$$\omega_R = \frac{1}{\sqrt{LC}}$$

By adjusting C, the resonant condition for a range of frequencies can be obtained.

Measurements of the electrical properties of a circuit using conventional instrumentation can be related to microstructural properties of the ferrite using a range of theoretical models, the choice of which depends upon high and low frequency response of the material.

3.4.18 Low Resistivity, High Permeability Ferrites

For relatively low resistivity, high permeability ferrites (e.g., Mn-Zn), the main cause of loss is by **eddy currents** within the material. These currents arise from the change in magnetic flux within the ferrite as the field reverses upon the application of an AC signal to the outer coil. The magnitude of the induced back emf depends upon the frequency. This current, acting through the resistance of the ferrite, leads to the generation of heat and hence loss.

Electrically, the rms power dissipation per unit volume, for a cylindrical ferrite of radius a and **resistivity** ρ is given by:

$$P_{rms} = \frac{a^2 \omega^2 B_{max}^2}{16\rho}$$

Magnetically, the power dissipation is expressed in terms of the **permeability**:

$$P_{rms} = \frac{1}{2} \mu_r \mu_o H_{max}^2 \omega \tan \delta$$

Thus: $P_{rms} = \frac{1}{2} \mu_r \mu_o H_{max}^2 \omega \tan \delta = \frac{a^2 \omega^2 B_{max}^2}{16\rho}$

Or: $\tan \delta = \dfrac{\mu_r \mu_o A f}{4\rho}$

where f is the frequency in Hz, A is the cross–sectional area of the core and ρ is the resistivity of the core. More usually, this is expressed as:

$$\frac{\tan \delta}{\mu_r} = \frac{\mu_o A f}{4\rho}$$

since this allows ferrite materials to be compared against one another without consideration of the geometry of the specimen.

The **relative permeability** is found from the ratio of the measured inductance to the air inductance L_o found from:

$$\mu_r = \frac{L}{L_o} \quad \text{where} \quad L_0 = \frac{\mu_o N^2 A}{l} \quad \begin{array}{l} \mu_o = 4\pi \times 10^{-7}\,\mathrm{H\,m^{-1}} \\ l = \text{length of coil} \\ A = \text{cross-sectional area of coil} \end{array}$$

In practice, allowance has to be made for the **stray capacitance** of the coil since this affects the measured values of Q and hence L. As well, the model does not include losses associated with changes in magnetisation of the material induced by domain wall movement. Despite these deficiencies, the essential feature is that the power loss within the ferrite increases linearly with frequency and can be measured experimentally from tan δ.

3.4.19 High Resistivity, Low Permeability Ferrites

In a high resistivity ferrite, such as Ni-Zn, eddy current losses are not significant even at very high frequencies (MHz). The primary loss mechanism is **domain wall relaxation**. At low frequencies, the domain walls move in response to the applied magnetic field and the power loss is low. At high frequencies, the walls cannot respond fast enough and the power loss increases, but, unlike the case of a low resistivity ferrite, reaches a saturation value at very high frequencies.

The **domain wall relaxation time** τ is a measure of the response time of the domain walls and is a constant for a particular material. When the period of the applied field becomes similar to the relaxation time, the ability of the walls to follow the field reversals in reduced.

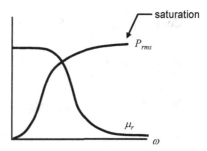

The **relative permeability** and **loss factor** depend on the frequency:

$$\mu_r = \mu_{ro}\frac{1}{\left(1+\omega^2\tau^2\right)}$$

$$\tan\delta = \omega\tau$$

At high frequencies, the permeability decreases because, for a given applied field strength, the domain walls do not move very far. At very high frequencies, the domain walls remain essentially stationary and the relative permeability approaches zero. At low frequencies, the permeability changes very slowly with increasing frequency and is almost constant.

In practice, at moderately high frequencies, the **loss factor**, $\tan\delta$, includes a contribution from eddy current losses within the surrounding copper windings and an additional term is required:

$$\tan\delta = \omega\tau + \frac{A}{\sqrt{f}}$$

Thus, to measure the relaxation time, $f^{1/2}\tan\delta$ is plotted against $f^{3/2}$ in the region where μ_r is fairly constant and the relaxation time τ is determined from the slope. A typical relaxation time for an Ni-Zn ferrite is on the order of about 40 nsec and so losses within such a ferrite are not expected to occur until the applied field is in the GHz region.

3.4.20 Example Properties of Ferrites

Ferrite	Ionic distribution		Saturation moment /molecule
	A	B	M_s
$NiFe_2O_4$	Fe^{3+}	$Fe^{3+}Ni^{2+}$	$2\mu_B$
$CoFe_2O_4$	$Fe_{0.8}^{3+}Co_{0.2}^{2+}$	$Fe_{1.2}^{3+}Co_{0.8}^{2+}$	$3.8\mu_B$
$Zn_{0.3}Ni_{0.7}Fe_2O_4$	$Zn_{0.3}Fe_{0.7}^{3+}$	$Ni_{0.7}^{2+}Fe_{1.3}^{3+}$	$4\mu_B$
$MgFe_2O_4$	$Fe_{0.9}^{3+}Mg_{0.1}^{2+}$	$Fe_{1.1}^{3+}Mg_{0.9}^{2+}$	1

Ferrites have 8 molecular units per unit cell with unit cell parameters:

	$\overset{o}{A}$
$NiFe_2O_4$	8.34
$CoFe_2O_4$	8.35
$Zn_{0.3}Ni_{0.7}Fe_2O_4$	8.37

For $NiFe_2O_4$, we have 4 unpaired 3d electrons, and so the saturation moment per unit volume is:

$$M_s = N\mu_M$$
$$= \frac{(8)(2)\mu_B}{(8.34\times10^{-10})^3}$$
$$= 2.56\times10^5 \, A \, m^{-1}$$

With no external field, we have $H = 0$ and the saturation flux density is:

$$B = \mu_o H + \mu_o M$$
$$= \mu_o M$$
$$= 4\pi\times10^{-7}(2.56\times10^5)$$
$$= 0.321 \, T$$

3.5 Superconductivity

Summary

$$B = 0$$

Conditions inside a superconductor

$$\chi = -1$$

$$M = -H$$

$$H_C = H_o\left(1 - \left(\frac{T}{T_C}\right)^2\right)$$

Critical field for Type I superconductor

$$I_C = 2\pi H_C$$

Maximum current for Type I superconductor

$$B = B_o e^{-\frac{x}{\lambda}}$$

$$\lambda = \left(\frac{m}{\mu_o n q_e^2}\right)^{\frac{1}{2}}$$

London penetration depth

$$\Delta\Phi = \frac{h}{2q_e}$$

Quantum of magnetic flux

3.5.1 Superconductivity

The superconducting state of matter has two independent physical properties of interest. **Superconductors**:

- have zero resistivity or infinite conductivity (Onnes, 1911).
- expel all magnetic fields from their interior and so are perfectly diamagnetic (Meissner, 1933).

The superconducting state of matter occurs in materials at very low temperatures. Only certain materials show the effect. Superconductivity disappears if:

- the temperature is raised above a critical temperature T_C,
- or if an applied magnetic field becomes too large,
- or if the current associated with the superconducting state becomes too large.

The infinite conductivity of a superconductor can be demonstrated by preparing a ring of superconducting material, and then withdrawing a magnet from the interior to begin the flow of current.

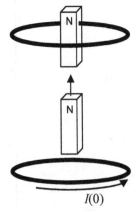

$I = 0$

When the magnet is pulled out, current $I(0)$ is induced in the ring. In a normal conductor, the current decays according to the inductance and resistance of the ring with a time constant L/R. In a superconductor, the current does not decay and remains circulating.

$I(0)$

If a superconductor is placed in a magnetic field, the field lines only penetrate to a very small depth in the surface and are expelled from the interior. This is the **Meissner effect**.

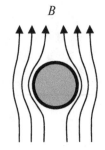

B

Inside the superconductor, we have:

$B = \mu_o(1 + \chi)H$

$B = 0$

$\chi = -1$

$M = -H$

Perfect diamagnetism

3.5.2 Quantised Magnetic Flux

When a current is established in a superconducting ring, say, by the
application of a uniform external magnetic field, it is found that the current
circulates within the ring. In fact, there are two currents, one circulating in
one direction on the outer surface, and one circulating on the inside surface
in the opposite direction. The magnetic fields associated with these two
currents cancel each other out, thus shielding the interior of the material
from the net field. This is the Meissner effect.

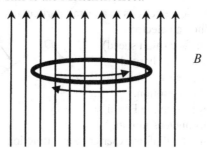

B

When the external field is removed, the current on the external surface
disappears and the one on the interior surface remains. This current can
circulate indefinitely. The magnetic field associated with the interior
surface current remains and loops around the conductor in the usual way
(right hand rule).

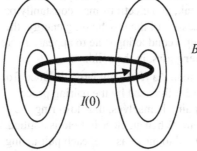

B

$I(0)$

It is found that if the strength of the applied field B is changed, then the
flux associated with $B(0)$ changes in discrete steps – that is, magnetic flux
is quantised. The quantum of magnetic flux is:

$$\Delta \Phi = \frac{h}{2q_e}$$

3.5.3 BCS Theory

As mentioned previously, the presence of the periodic crystal potential only provides a different modulation to the wave function for each value of k. Resistance is a consequence of the existence of imperfections in the crystal lattice that disrupt the periodicity of the crystal potential and the presence of thermally induced lattice vibrations (**phonons**). Ironically, it is the presence of lattice vibrations in special circumstances that accounts for the phenomenon of superconductivity.

Consider the conduction electrons in a metal, particularly those near the **Fermi surface**. When an electron e_1 near the Fermi surface (high speed) passes near to a positively charged nucleus, the nucleus feels a Coulomb attraction and is set into motion. Another electron e_2 near to the Fermi surface may see this nucleus moving

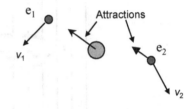

away from itself and it will thus experience an attraction. The net effect is that the nucleus has screened the natural Coulomb repulsion between e_1 and e_2 and indeed, e_2 now appears to be attracted to e_1. Technically, e_1 has emitted a phonon which is absorbed by e_2.

This attraction only occurs under certain circumstances: when the electrons have opposite spins and are travelling in opposite directions ($k_1 = -k_2$). The binding is fairly weak, and electrons may constantly pair and break up with other electrons. Because of the low binding energy, the temperature T has to be low for the superconducting state to appear. There may be many such pairs, called **Cooper pairs**, in the solid.

Cooper pairs are situated a relatively large distance from each other. Other nearby electrons are influenced by the pairing process with the result that they tend to form pairs themselves, thus lowering the overall potential energy of the system. When an electric field is applied, all the ordered pairs of electrons tend to move as one, each pair acting as a negatively charged particle of $2q_e$. It is the correlated motion of Cooper pairs that is responsible for the superconducting state. In essence, lattice vibrations, which tend to inhibit conduction for ordinary conductors, provide a mechanism for ordered, unhindered motion of charge carriers in a superconductor.

3.5.4 Energy Gap

The **binding energy** associated with a Cooper pair is of the order of $3kT_C$ where T_C is the critical temperature for the material. To obtain the superconducting condition, electrons near the Fermi surface form **Cooper pairs**. The energy required to do this is the binding energy of the pairing, or phonon interaction. In effect, the binding energy manifests itself as an **energy gap** in the energy spectrum, centred at the **Fermi level**.

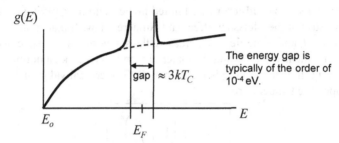

The energy gap is typically of the order of 10^{-4} eV.

At 0K, all the states below the gap are fully occupied. In the superconducting state, those in the states near the gap form Cooper pairs. As the temperature is raised, the gap narrows, disappearing completely at $T = T_C$ where the density of states is the shape of a normal conductor and pairing is destroyed by random thermal excitations. Lowering the temperature allows pairing to be re-established and the gap to reform. The process is reversible.

Note that Cooper pairs occur at the opposite sides of the Fermi surface because $k_1 = -k_2$.

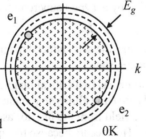

As the temperature is raised, the gap narrows and eventually closes and pairs no longer form.

3.5.5 Type I and Type II Superconductors

Experiments show that the superconducting state is destroyed if the temperature is raised above a **critical temperature** T_C or if an applied magnetic field becomes too large, or if the current in the superconducting state becomes too large.

If the temperature is raised, then lattice vibrations associated with temperature prevent any long range pairing required for superconductivity.

Type I superconductors are limited in their current-carrying capacity because of the Meissner effect and Ampere's law. If placed in a magnetic field, H, and if the field is increased, the superconducting state disappears at a critical value H_C and the material then behaves like a normal conductor even if the temperature is lower than T_C. The critical value of H depends upon the temperature:

$$H_C = H_0\left(1 - \left(\frac{T}{T_C}\right)^2\right)$$

where H_0 is the critical field at 0 K. As the temperature is raised from 0 K to T_C, the value of H_c decreases – that is, the lower the temperature, the greater the critical field.

Passage of a current in a conductor is accompanied by a self-induced magnetic field. In a superconductor, the current and the associated magnetic field are confined to the surface layer of the solid. The maximum current that can be carried by a Type I superconductor of radius r is:

$$I_C = 2\pi r H_C$$

At currents greater than this, the self-induced magnetic field is sufficient to destroy the superconducting state.

A **Type II superconductor** has two critical magnetic field intensities, H_{C1} and H_{C2}. Such materials have the ability to remain superconducting while not excluding a magnetic field from the interior as long as it is between these two critical levels. This occurs because the applied magnetic field creates an array of cores or vortices of material which enter the normal state while leaving the material outside in the superconducting state. Eventually, the flux lines associated with the vortices overlap and the entire material enters the normal state. Type II superconductors are able to carry larger amounts of current compared to Type I because H_{C2} can be hundreds of times larger than H_C.

3.5.6 London Penetration Depth

The penetration of a static magnetic induction B into a superconductor can be expressed as an exponential:

$$B = B_o e^{-\frac{x}{\lambda}}$$

$$\lambda = \left(\frac{m}{\mu_o n q_e^{\,2}} \right)^{\frac{1}{2}}$$

where λ is a characteristic length called the **London penetration depth**. m is the mass of an electron, q_e is the charge on an electron and n is the number density of electrons (i.e., superelectrons).

A representative estimate of the penetration depth λ can be obtained by setting the number density $n = 1 \times 10^{29}$ m^{-3}:

$$\lambda = \left(\frac{m}{\mu_o n q_e^{\,2}} \right)^{\frac{1}{2}}$$

$$= \left(\frac{\left(9.1 \times 10^{-31}\right)}{4\pi \times 10^{-7} \left(1 \times 10^{29}\right)\left(1.6 \times 10^{-19}\right)^2} \right)^{\frac{1}{2}}$$

$$= 1.68 \times 10^{-8}\, m$$

$$= 16.8\, nm$$

This is the London penetration depth, and represents the extent of the magnetic induction inside the superconductor. It also implies that the current that flows within a superconductor does so very near to the surface of the superconductor.

3.5.7 *B-H* and *M-H* Loops

Type I superconductor

Type II superconductor

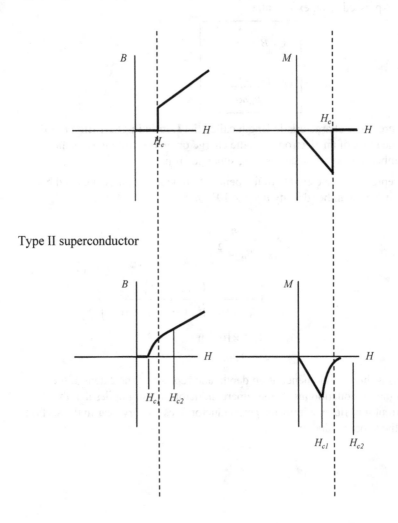

3.5.8 SQUID

Quantum mechanics predicts that wave function for an electron may tunnel through a barrier even though the barrier potential may be greater than that of the electron energy.

In 1962, Josephson predicted that if a superconducting material were to be placed either side of an insulating junction, then Cooper pairs may tunnel through the junction. Experimentally, this manifests itself as a current across the junction, even when there is no external voltage applied. This is the **Josephson effect**.

The phase of the wave function of the Cooper pairs in the two superconductors will not be equal. The magnitude of the current observed flowing across the junction depends upon the phase difference and is limited to the critical current I_C.

When an external DC voltage is applied across the junction, an alternating *current* appears across the junction with amplitude I_C. The frequency of the current depends upon the magnitude of the applied DC voltage V.

$$\omega = \frac{2eV}{\hbar}$$

A typical applied voltage is on the order of a few mV, putting the Josephson frequency in the GHz range.

One application of the Josephson effect is in the construction of a sensitive magnetometer: **SQUID**. In this device, a ring of superconducting material is made with two junctions inserted.

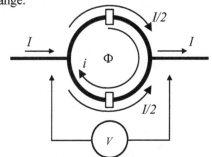

The characteristics of the junction depend very greatly on the magnitude of the magnetic flux through the ring. Each time the B field increases or decreases by ½ a magnetic quantum, the direction of the current i within the ring undergoes a reversal of direction which either adds or subtracts from the current I in the different halves of the ring. B fields on the order of 10^{-14} T can be detected using this apparatus.

Index

Printed in the United States
by Baker & Taylor Publisher Services

Printed in the United States
by Baker & Taylor Publisher Services